BEST TEAM EVER

BEST TEAM EVER

THE SURPRISING SCIENCE OF HIGH-PERFORMING TEAMS

DAVID BURKUS

TWINBOLT

Best Team Ever: The Surprising Science of High-Performing Teams

Hardcover ISBN: 978-1-5445-4175-4

Paperback ISBN: 978-1-5445-4174-7

Ebook ISBN: 978-1-5445-4176-1

Audiobook ISBN: 978-1-5445-4177-8

To those who do work that matters,
which is all of you.

CONTENTS

INTRODUCTION

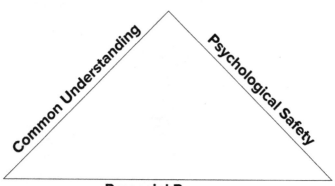

USA CURLING HAD A PROBLEM.

Their Olympic teams were failures. Repeat failures, actually.

In 2010, Team USA finished the Olympic curling competition—known in curling as a *bonspiel*—in exactly last place. Their performance was so bad that their team captain—known in curling as a *skipper* or *skip*—had to be benched in an attempt to stop their failing performance. That skip, John Schuster, would become well-known shortly after his total failure at the 2010 Games. His failure was so bad that an entry for "Schuster" was once added to Urban Dictionary, the website for pop culture words and phrases, the Wikipedia of slang. At the time, the definitions read: "A verb meaning to fail to meet expectations, particularly at a moment critical for success or even slightly respectable results. Slightly derogatory, indicating the type of disapproval that can only be backed by the weight of a nation's scorn."

Team USA's performance in the 2014 Olympic Games was slightly better but equally disappointing. This time, somehow again led by John Schuster, Team USA finished the bonspiel in second to last place.

The repeated failures were seen as Schuster's fault. The problem of fixing things fell on USA Curling, the governing body for all curling in the United States. USA Curling wanted to assemble a better team—known in curling as a *rink*—to take the ice, which is known in curling as a *sheet* (these names are meant to avoid confusion but actually create more of it), before 2018's next Olympic *bonspiel*. To accomplish that, USA Curling launched what they called a "high-performance program" designed to find promising talent and develop it over years into a *rink* of chosen curlers led by a superior *skip*… Okay—actually, let's be done with the curling lingo.

For obvious reasons, John Schuster and the other failures from the 2014 games were not invited to participate. But that didn't stop any of them.

Schuster assembled a team of his own. He partnered with John Landsteiner and Matt Hamilton, both of whom had also been rejected by USA Curling; and with Tyler George, who hadn't been rejected, but mostly because he'd never bothered to try out. They trained together and competed together, determined to work their way back onto the curling's world *sheet*. (Sorry, I couldn't resist.)

They weren't well funded—each had real jobs they had to maintain just to pay for their equipment and their bills—but little by little, the group of four, known in curling circles as "Team Reject," started to attract some attention. In 2015, Team Reject beat both of the teams enrolled in the high-performance program at the national curling championships. Their win obligated USA Curling to send Team Reject to the world championships, instead of sending their chosen teams. Schuster and his team finished in fifth place. It was not ideal, but it was way better than last time—and good enough for Team Reject to be un-rejected and invited to also join the high-performance program.

In the 2016 world championships, they did even better and won a bronze medal for the United States. In 2017, they failed to medal but did finish well enough to earn an automatic bid to the Olympic trials. At those trials, Team Reject had to face off against all of the teams hand-selected by USA Curling—the supposed best of the best—and Team Reject bested each of them.

Their victory led to an even bigger problem for USA Curling.

They now had to send a team they saw as failures and rejects to the Olympic Games one more time. The move was criticized by just about everyone—and no one was left uncriticized. USA Curling and the hand-selected teams were criticized for failing to put down the rebellion from Team Reject. Team Reject was criticized for being…well, rejects. And John Schuster was criticized for even wanting to lead a team of such proven failures. Some critics went so far as to say that USA Curling should have neglected to send a team at all, and thus spared the United States from another Olympic embarrassment.

And in the first few rounds of Olympic competition, it looked as if those critics may have been right.

Schuster and the team were failing—again. They lost four of the first six games, almost leading to their removal from competition. In order to have a chance at a medal,

Team Reject would have to win every single game left in their schedule of round robin play.

And that's just what they did.

"We've played our best when our backs were up against the wall. We took it to another level this week," said Tyler George, aptly describing not only their performance during the second half of the Olympics, but also the last four years of their curling careers. "Usually, we're fighting and scrapping to get into the playoffs, but for five days we were the best team in the world, and we did it at the right time."[1]

In the medal rounds, Team Reject—now Team USA—went on to defeat three-time defending Olympic champions Team Canada and then played Team Sweden for the gold medal. Faced with needing to win or go home, Team USA curled a nearly perfect game. They were so good, in fact, that after Schuster threw a pristine stone, the captain for Team Sweden conceded the match, handing them the gold medal.

Team Reject had become American curling's best team ever. Their dominance could no longer be denied—even though USA Curling tried to deny it for years. In the 2022 Winter Olympic Games, three of the four members of Team Reject were chosen to represent the United States once again. And, in a twist of fate, the fourth man on their rink was Chris Plys, who had replaced Schuster after his 2010 benching. This next iteration of Team Reject finished in a respectable fourth place. Schuster and company didn't finish on the podium, but even before the competition began, they were honored for their career success. Schuster was chosen to lead Team USA during the opening ceremonies as a flag bearer.

The man nobody had wanted to lead USA Curling found himself leading all of Team USA.

Why Teams Win

This is not a book about Team Reject.

Well, those first few pages were.

This is really a book about USA Curling's problem. It's the same problem that's nearly universal in sports, business, and everywhere else teamwork is necessary. It's a problem you've likely had in the past and definitely will have in the future. It's the problem of explaining why teams win. Or, better said, it's the problem of building a team with the best chance of winning.

Humans were designed to work in teams. The archeological evidence suggests that the earliest humans lived and hunted in tribes. Even as hunter-gatherers gave way to farmers, humans still preferred to work in teams. Farming together actually allowed them to expand their teams and build whole cities. We can still see the imprints of those early humans in our anatomy today. Our brains, facial muscles, vocal cords, and so much more were developed to enhance our ability to collaborate.

And enhance our collaboration abilities we have.

Today, the entire world runs on teams. Jobs that could have been solitary at one time or another are done more efficiently and at higher levels of quality because we work in teams. Teams build our houses and manufacture the appliances we put inside of them. Teams perform surgeries and develop our medicines. Teams teach our children. Teams protect our society. Teams run our worship services. Teams write and edit the books we read and the software we work on. We work in teams—and many of us work on multiple teams. The number of teams we form, along with the size of those teams, has increased exponentially since our ancestors first formed teams to chase down prey.

Accordingly, the importance of building teams that perform well together has also increased. Not just any teams—but teams that share a great culture.

You've probably felt what it's like to be on a high-performing team. You feel energized. Your brain is on fire with great ideas, and conversations with the team spur even more. You finish work each day with more energy than when you started. Unfortunately, you've probably also felt what it's like to be on a team with a broken culture. You end each day drained. You feel let down after every meeting and wonder if it's worth it to continue.

And, if you're like most, you've probably felt the impact of a positive or negative team culture even more strongly over the last few years. As more collaboration has occurred in teams, team culture has become the dominant influence on our experience of work. During the COVID-19 pandemic, many of our interactions with coworkers were limited to the dozen or so people serving on the one or two teams we worked most closely with. If you were asked to describe your company's culture, odds are you would share much more about your team's culture than about the company as a whole.

We know we need to build the best team we can, but many of us still wonder how.

Fortunately, decades of evidence—both in the laboratories of research universities and in the hallways of great organizations—has solved a lot of the mystery surrounding high-performing teams. While each study, each paper, and each researcher explores teams from a slightly different approach, all share a few core findings.

All assert that not all teams perform equally—and that the best-performing teams operate at a dramatically higher level than do everyday groups.

Talent doesn't make the team. The team makes the talent.

All assert that the culture of the team is more important than who is on it. Team culture—the collective values, beliefs, behaviors, and ways of working a member's share—has an outsized effect on the results a team achieves. You can't recruit your way out of a lousy culture. Talent doesn't make the team. The team makes the talent.

And, while they may use different terms to describe them, all assert that high-performing team cultures share three fundamental elements: Common Understanding, Psychological Safety, and Prosocial Purpose.

Common understanding happens when team members understand the team's expertise, assigned tasks, context, and preferences. *Psychological Safety* builds as team members learn to feel safe to express themselves, take risks, and disagree respectfully. And *Prosocial Purpose* motivates teams by showing them the meaningful contribution they make toward work that impacts others.

These three elements provide teams with a common set of norms and behaviors that guide their collaboration and improve their performance. As an organizational psychologist, a former business school professor, and now a writer and keynote speaker, I've been studying and teaching how these three elements—and team culture as a whole—affect team performance for nearly two decades.

This book is designed to be part training manual and part expedition guide through each element. Across three sections, we'll discuss the research that explains why each element is so important. I'll provide practical ways leaders (or even influential team members) can begin working on each element. Along the way, we'll explore a few powerful examples of these elements at work in high-performing teams—from astronauts to entrepreneurs, from healthcare heroes to fast food workers, and from corporate executives to Olympic gymnasts. We'll even learn a bit from the greatest baseball team you've never heard of.

Common Understanding, Psychological Safety, and Prosocial Purpose.

Build these, and you've begun to build your best team ever.

PART ONE

COMMON UNDERSTANDING

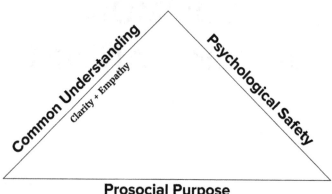

CHRIS HADFIELD IS THE MOST FAMOUS LIVING ASTRONAUT. You may have watched one of his hundreds of short online videos about what it's like to brush your teeth or cut your hair in space. Or, you may be one of the over fifty million people who watched the Canadian famously cover David Bowie's "Space Oddity" while floating around the International Space Station. (If you haven't, you really should. You can put this book down—it'll still be here in five minutes when you're done watching).

Hadfield's videos showed what it was like to live for long periods of time in space, and allowed those of us stuck on earth to empathize with the astronauts in a new way.

But inside the space community, Hadfield is known for so much more. He's known for completing three trips to space over seventeen years, and for holding just about every leadership position at NASA when he was on the ground. He's one of only approximately two hundred people who have ever walked in space, and he's one of the best commanders of a mission to the International Space Station—ever.[2]

Hadfield was named commander of the International Space Station's Expedition 35, where he and four other astronauts lived for five long months. From the beginning, Hadfield knew how much of a challenge those months would be—and how much of his success would come down to preparation. Hadfield launched into space alongside American Tom Marshburn and Russian Roman

Romanenko. They were joined months later by American Chris Cassidy and two more Russians: Pavel Vinogradov and Aleksander Misurkin. (ISS missions overlap their crews to ensure operations run smoothly).

Hadfield led a team with three distinct cultures: Russian, Canadian, and American.

And three distinct languages: Russian, English...and American.

So, while much of his preparation involved training to live and work at the space station, he focused much of the team's time on living and working *together*, as well.

Hadfield has a reputation for his diligent preparation, always taking extra courses and becoming certified to operate machines and spacecraft he may never actually operate "just in case" he may need to. And NASA training diligently prepares astronauts, as well. There are hours of classroom training and simulation—which even includes scuba diving in a giant pool with a submerged replica space station. There are checklists upon checklists. There is an ever-growing book called *Flight Rules*, which offers guidelines for every conceivable situation—and, when the inconceivable occurs, new guidelines are added. "*Flight Rules* are the hard-earned bodies of knowledge recorded in manuals that list, step-by-step, what to do if X occurs and why," Hadfield explained. "They are extremely detailed, scenario-specific standard operating procedures."[3]

When you're in space, ground control has your workday planned in five-minute increments, sometimes including what you're supposed to eat and drink. Teams are given a lot of clarity around what, how, and when to perform—but not as much clarity on who they're performing with and how to get along effectively. "The longer the flight, the more important personalities become," Hadfield reflected. "If the three of you don't get along on Earth, you're even less likely to be able to tolerate each other after a few months without the benefit of showers. Or Scotch."[4] Hadfield knew that clarifying expectations and accepting personality differences would be key to a successful mission.

So, Hadfield focused his authority on gaining that clarity for the team. He spent time living and working in both the United States and Russia. He learned Russian while living in state housing (forgoing the American-style houses NASA had built on base in order to better understand the culture). He also brought the team together to understand each other as much as they could. They talked about each other's preferences. They met each other's families. Hadfield even had them role-play certain scenarios to uncover how they'd respond, including uncomfortable situations like the loss of a loved one while in space.

And that dual understanding—clarity of tasks but also clarity about each other—made the mission a resounding success. Despite being confined together in a tin can circling the earth for five months, the team never once had a heated argument.

There were plenty of surprises that would test their resolve. Three months into the mission and only thirteen days after Hadfield's team took command of the Space Station, one of the astronauts, Tom Marshburn, was notified that his mother had died.[5] But the team was prepared and could help him mourn. With just a few days left in the mission, an ammonia tank on the station began leaking and needed to be fixed with a spacewalk. The team had just one day to prepare for what normally takes weeks, but they successfully stopped the leak.

The strong culture of the team may even be one of the reasons Hadfield's videos performed so well—you can't lip-sync David Bowie while floating around and playing guitar without someone to film you (and someone else to keep you from smashing into things). The level of collaboration Hadfield built enabled them to do more than just the spacewalks and scientific experiments they were tasked to do. Ironically, their extra projects were what revealed to most of us groundlings the humanity behind space travel. Had Hadfield not focused on the humanity of his team, it's possible we'd never have had as clear an understanding of "what they're doing up there" as we do.

Hadfield's leadership demonstrates the power of building *common understanding* on a team, the first element of a high-performing team culture. Common understanding happens when team members understand the team's expertise, assigned tasks, context, and preferences.

Teams work best when they understand how each individual works best.

Different team members have different knowledge, skills, and abilities—not to mention different tasks. Team members need to know how their experience and their assignments fit into the larger whole. They need to know who knows what, but also how to approach each person with requests for help—or offers of help. Teams work best when they understand how each individual works best.

When team members don't have common understanding, they begin to doubt whether their teammates are competent. They doubt whether their team members will deliver on shared goals. And those doubts decrease motivation and engagement across the team.

When team members do have common understanding, it makes collaboration faster and makes coordinating roles easier. Beyond trusting each member to deliver, high-performing teams with common understanding are able to anticipate each other's actions and respond quickly. Common understanding builds the playing field that allows this level of performance.

Common understanding even raises the level of intelligence of the team as a whole.

In 2010, a team of researchers led by Anita Williams Wooley sought to study how the inner workings and behaviors of teams affected their end performance. They were especially interested in how the expertise of individual team members affected the overall performance of the team. To figure this out, they recruited 152 teams and gave each a series of different assignments that required collaboration—meaning that teams had to work together the entire time, not just bring their answers together at the end. The tasks ranged from creative challenges like idea generation to decision-making challenges that involved planning ahead by coordinating tasks. In total, teams were given three different tasks that involved three different modes of thinking. Some teams performed well on all three assignments, and some teams fell apart during all three. Interestingly, how well a team performed on one task predicted how well they'd perform on the others. While the composition of the team and individual skill sets might randomly favor one team for one particular task, it didn't seem to matter.

In addition, the average intelligence of the group didn't seem to predict results, either. The researchers gave each participant an IQ test beforehand and found no correlation once the tasks were completed. It wasn't what any individual member brought to the team; it was how the team acted when together that mattered. Instead of average individual intelligence, the researchers argued that these teams developed a *collective intelligence.* Once they developed norms of behavior and understood what each member could offer, coordinating and executing the task became easy—no matter what the task.[6]

In other words, how much the team developed common understanding affected how collectively intelligent they became—and that affected performance more than anything else.

One reason is that common understanding and collective intelligence allow teams to collaboratively plan their work. This is something some of the researchers discovered before. Taking the time to collaboratively plan tasks and think through how those tasks fit together resulted in lesser-qualified teams achieving more success than more-qualified teams in prior studies.[7] But it wasn't just making plans based on what was overtly assigned. It also required finding information beneath the surface.

Every team member takes for granted certain assumptions around norms, values, and behavior. But great team performance involves discovering and revealing those taken-for-granted assumptions to avoid disaster. In this way, common understanding on a team also helps to explain where teams suffer breakdowns—and how to avoid them. According to the United States National Transportation Safety Board, around 73 percent of flight incidents happen on the first day of an airplane crew flying together, and 44 percent of those occur on the crew's very first flight—when clarity of roles and personalities is at its lowest. In a study conducted by NASA, fatigued crews with a history of working together (having therefore developed some common understanding) made half as many errors as fully rested teams who had not flown together before.[8]

Developing common understanding involves making sure team members have role clarity and dependability, but it also involves building empathy among members.

Clarity and Empathy.

When teams understand not just their roles and responsibilities, but also the emotions and desires of their teammates, they operate at a higher level. In this section, we'll examine common understanding in two chapters by focusing on clarity and empathy in turn. First, we'll outline how to build the clarity on a team that allows for collaborative planning. Second, we'll examine how building empathy allows teams to raise their level of collective intelligence. And we'll do this by reviewing the research, but also by explaining what two seemingly opposite worlds have in common: fast-moving gymnasts, and even faster food.

CHAPTER ONE

CLARITY

WHEN YOU THINK OF TEAMS THAT OPERATE WITH precision, even under strenuous circumstances, you might think of Formula One racing crews, world-class trauma surgeons, or championship sports teams. You generally don't think of fast-food workers. Even if you did, you probably don't think of Pal's Sudden Service—because unless you live in the Tri-Cities Tennessee region, you've likely never heard of Pal's. But if you have, then you know that the precision and clarity with which their teams serve you up a Sauceburger and Frenchie Fries rivals that of racing crews and trauma surgeons.

Pal's was started in 1956 by Fred "Pal" Barger after he met and was inspired by McDonald's CEO Ray Kroc. But Barger didn't just want to mimic McDonald's; he wanted to improve on it. (The chain's name is a hint at Barger's intentions. It's not fast food; it's sudden service—because, in Barger's mind, the only thing faster than fast is *sudden*.)

Their service is indeed sudden—or at least faster than fast. Pal's measures their service time as "wheels stop to wheels go." After their order is taken, not through a speaker but by a real person (more on that later), customers stop at the pick-up window for an average of only eighteen seconds before they drive off with their freshly-handmade order. That's less time than it takes to sing "Happy Birthday" to each other in the car. The same service at McDonald's and other well-known fast-food chains takes, on average, about three times as long. That's a lot of "Happy birthday to you."

And their service is not just sudden. It's also close to perfect. Pal's employees only make a mistake once in every 3,600 orders. That's ten times better than the average fast-food joint. Regular customers don't even check their bag after picking it up. They know it's perfect.

Pal's even won the 2001 Malcolm Baldridge National Quality Award, becoming the first restaurant ever to win the nation's premier award for performance excellence. (And it's worth noting that the second restaurant ever to win the award sent their leadership on dozens of trips to learn from trainers at Pal's). Pal's is well known in the industry for near-perfection delivered nearly instantly. In 2000, after so many organizations asked to visit and learn from Pal's, the company started the Business Excellence Institute (now called the McClaskey Excellence Institute, named after David McClaskey, the partner who helped Pal's win the Malcolm Baldridge Award).

One way Pal's achieves excellence is through design. Their menu is simple: burgers, sandwiches, fries, and drinks. And it doesn't change—new or seasonal items would require more training or result in more mistakes. The majority of locations are drive-through only. Customers give their order to a human through a window, not a speaker, which cuts down on miscommunications that lead to mistakes.

But the biggest contributor to Pal's speed and precision is the level of role clarity that can only come from rigorous training. Before a new hire is allowed to work in the kitchen, they have already completed over 120 hours of training. Managers receive up to 800 hours of training. The goal is for every employee in the store to be trained in every process at every station. It's as if someone at Pal's read NASA's *Flight Rules* and thought, "Let's apply this specificity to making burgers."

Most of the training is delivered by computer programs and testing, but trainers and the store's owner-operators also give one-on-one coaching. In fact, leaders at all levels give training. One company policy is that all leaders must spend 10 percent of their time helping employees develop—including the CEO, Thom Crosby. "The core part of every manager's job is to be an educator," Crosby has said.[9]

Pal's employees aren't just trained and set loose; they're retested often—or, as Crosby likes to put it, "recalibrated." Every day, at every store, a computer randomly assigns a few employees to a pop quiz that will retest them on one of their jobs. The quizzes are handed to the employees at the beginning of their shift, and they have until the shift ends to turn in their answers. Cheating is encouraged; if an employee doesn't know an answer, they can ask other team members. The goal isn't to punish employees; it's to make sure they've been given enough clarity to succeed. "If people aren't doing something right," Crosby asserted, "that's not a problem with them. It's a problem with the training."[10]

That mentality is what keeps Pal's at peak performance. It also keeps employees from feeling micromanaged, and keeps the locations from feeling like factories. "When Pal's staff members look to the left and then to the right, they see everyone operating at a high level, and they follow suit," said David Jones, a senior trainer for Pal's at the McClaskey Institute.[11]

And Pal's team members know their coworkers at a given location are operating with excellence, because they've also been trained in their jobs. Pal's locations use fifteen different stations, and each employee is trained in more than one station. Some are trained in all fifteen. Moreover, it's not until employees arrive for their shift that they learn which stations they'll be working on that day. The novelty keeps team members engaged, but it also helps everyone see how their tasks fit into the larger process of serving customers at near-perfection. Pal's employees even reinforce performance among each other through a "Caught Doing Good" program, where individuals can recognize and nominate a teammate who is performing particularly well that day for a bonus—something that wouldn't really be effective unless everyone knew what was expected of everyone else. As David Jones described, "Life becomes easier when there is one set of rules followed by everyone."[12]

And life is good for just about everyone at Pal's. Turnover among frontline employees is around 32 percent—which is about a third of industry average. (If you're doing the math in your head, you figured out that the fast-food industry averages 100 percent turnover—meaning the entire front-line staff at your nearby burger chain quits every year.) Pal's turnover at the manager level is just 1.4 percent. All of that leads to a highly successful company. Pal's customers visit their local store about three times per week, compared to three times per month at a typical McDonald's. And Pal's locations average $1,800 in sales per square foot, compared to only $650 per square foot at McDonald's.

Fast food is an industry built on standardization and clarity. But Pal's Sudden Service proves that designing the perfect system is useless unless that clarity jumps from the training manual to the mind of every team member. When team members at Pal's are completely clear on their roles and how their roles fit into the entire system, they're not only more productive—they're also happier.

And why wouldn't they be? They're on the best fast-food team ever.

Getting Clear on Clarity

It may at first seem like Pal's takes clarity to a level totally uncalled for by a fast-food chain. And you could make that argument. But you can't argue with their results.

Pal's leaders understand that clarity is a gift you give team members that empowers them, and the whole team, to work better. When team members know what's expected of them, and they trust others to do it, they perform at a higher level. The more clarity is created, the better the team performs.

But wait. Doesn't making sure each person knows their specific role mean there's *less* reason to collaborate? Wouldn't it be better to paint a vivid picture of the end goal, and allow teams to decide for themselves how best to co-ordinate roles and responsibilities? Aren't leaders supposed to "cast a vision" and inspire the team to work toward it?

Those are certainly common assumptions, and ones shared in many popular writings and speeches.

Often, those writings and speeches share the same motivational quote, from *Little Prince* author Antoiné de Saint-Exupéry. "If you want to build a ship, don't drum up the men to gather wood, divide the work, and give orders. Instead, teach them to yearn for the vast and endless seas."

Doesn't that sound inspiring?

Well, all the yearning in the world isn't really going to help if there's no wood gathered. Unless certain members of the team understand buoyancy and waterproofing, your sailing will end before it starts. And Saint-Exupéry was an aviator, not a sailor. So, he can't help you much either.

As we'll see in Part Three, inspiration is important, but teams need specific guidance on what to do—and they need to stay up to date on who is doing what—or else that inspiration is all for naught.

It may seem that the more clarity you have, the less your team needs to communicate. It's assumed collaboration will suffer. But role clarity doesn't diminish collaboration. In fact, research suggests the opposite.

In 2007, Professors Lynda Gratton and Tamara Erickson were looking to discover the conditions that enabled the most collaborative teams. They studied fifty-five teams at fifteen multinational companies. The teams ranged in size. One of their chief findings was that collaboration improved when individual roles were clearly defined. When individuals felt they could do most of their work independently, the parts that required interdependence flowed more smoothly. "Without such clarity," Gratton and Erickson wrote of their findings, "team members are likely to waste too much energy negotiating roles or protecting turf, rather than focusing on the task."[13]

In instances where teams are leaderless or have been empowered by leaders to make decisions about roles, clarity matters just as much. How quickly those teams arrive at clarity of roles and create commitments to each other on tasks can be the difference between seeing a project to completion or seeing it stay in limbo.

This doesn't mean the best teams function like robots, with merely a list of predefined tasks. There may be times when the team's goals are somewhat ambiguous. In those cases, high-performing teams take the time to consider their options, arrive at a plan, and divide up the work to create clarity. The earlier in the process teams do this, the better they perform. Moreover, when circumstances change or force the team to pivot, the best teams get back together fast, get everyone up to speed on what's different, and get back after it.

Teams without strong clarity feel frustrating. Often, each team member starts feeling like their efforts are carrying the team and everyone else is letting them down. And yet, those other team members feel the same way in reverse. (And yes, I realize I've just described every project team you were ever on in college.) When you don't know what to expect from others, you're usually disappointed. Even though your teammates are working, you don't feel like you can depend on them.

In contrast, teams with strong clarity are exhilarating. Team members know exactly what is expected of them and by when, and that accountability motivates them to not only deliver upon expectations but exceed them. (After all, you can only exceed expectations when you know what those expectations are.) As teams continue to deliver, they grow in their level of trust and dependability. It's why team members at Pal's don't feel like cogs in a machine. Instead, they feel like players on a championship sports team.

Making Teams More Clear

Your team may feel frustrating or exhilarating depending on the current level of clarity, but there is always room to grow. Here are a few ways you can get started bringing more clarity and dependability to your team:

HOLD HUDDLES

The concept of huddles is borrowed and adapted from the daily scrum in Agile software development. (But "scrum" is a Rugby term, and—being an American—I have no idea how Rugby works.) In an Agile scrum, teams gather daily for a brief meeting and typically answer three questions individually:

- What did I just complete?
- What am I focused on next?
- What is blocking my progress?

While your team may not need these answers daily, the genius lies in the questions. "What did I just complete?" allows the team to report on what deliverables are complete and figure out what progress they made. "What am I focused on next?" allows the team to ensure that no critical tasks are being accidentally left undone. But

"What is blocking my progress?" may be the most powerful question of all. It allows team members to regularly signal where they need help, and it lets others volunteer their aid before a small obstacle turns into a big problem. The goal of the huddle is to help the team work out loud so everyone knows what's going on and any pivots that may be needed. The team can decide if daily meetings are necessary, or if weekly or some other cycle will suffice. If the team is spread out across multiple time zones or countries, then the huddle format can still work via team-wide emails or posts in a shared channel.

And it's important to keep focused on projects and objectives. If the meeting becomes a call for everyone to just read what's on their calendar, then it may be wise to change up the cycle or the questions to make the ritual more useful for your team.

COMMUNICATE IN BURSTS

As we saw from the research, just because a team collaborates well doesn't necessarily mean they're in constant communication. In fact, the opposite is true. As anyone whose calendar has ever been stacked with meeting after meeting can attest, too many meetings and communications often crowd out time to do actual work. High-performing teams' communication patterns are often marked with long periods of alone time for solo work, thorough research, or deep thinking. Then, in between those periods are bursts of communication where everyone returns to give status updates, solve problems, and make decisions. Depending on the work your team does, it is probably wise to adopt the same rhythm.

Start small. If your team and organization are used to calling a meeting about everything (including the problem of too many meetings), then moving immediately to restrict communication to small bursts will likely get rejected. Instead, maybe setting "heads up" and "heads down" times on the calendar would be a smaller intervention. Or, declaring a "no meeting day" (or even an afternoon) is a great first step. While you're at it, talk with the team about what norms around electronic communication need to be established—like no sending team-wide emails after 3 p.m. (And if that took your breath away, remember that replying to email and sending email are still two separate actions, even if your computer defaults to sending immediately.) Whatever you decide, once your team experiences a little success, consider extending the communication limits even further until you feel it's as infrequent as it can be without people feeling uninformed.

DRAFT TEAM CHARTERS

While huddles keep a team informed about who is working on what, how to work together is still an unanswered question. Every team has norms of behavior that guide interactions, but many teams develop these norms only after suffering dizzy spells of confusion when coworkers hold conflicting norms. Team Charters (sometimes called working agreements or social contracts) aim to alleviate this confusion. In drafting a charter, teams start with a list of questions around communication and collaboration that have gone unanswered, and then discuss each one to arrive at norms. These questions can be small or large, such as the following:

- What's a reasonable amount of time to wait for an email response before resending?
- How should we keep each other updated on our progress (huddles, emails, project management software)?
- How should we ask each other for help?
- What's the best way to make decisions (voting, consensus, yelling loudly until the last person loses her voice)?

These questions are just a starting point—and comic relief—but even this small list likely contains a question your team has never talked about. Before calling the meeting, pay attention to what's triggering miscommunication and conflict on your team. Don't attempt to resolve it with your own rule; just add it as a question to the list for when you hold your team charter meeting.

After each question has a suitable answer, capture each answer in a collaborative document that can be shared team-wide. Then agree on a time to reexamine how well these guidelines are working and what changes need to be made (or if the guidelines were even followed). The charter will be a living document—because it's not having the document that enhances collaboration, it's drafting and revising it together.

MAKE PRIORITIES CLEAR

Most teams juggle various tasks, including the task of keeping them all coordinated. In keeping that juggling going, priorities can get shuffled—especially when a lot of tasks suddenly become "urgent" even if no one has determined whether they're "important." But most high-performing teams start with clear priorities, and only a few of them. Unfortunately, the average team starts with a vague notion of priorities. Several years ago, *Inc.* magazine asked senior leaders at six hundred different companies to estimate how many of their employees knew the top priorities of the organization. The average senior leader estimated 64 percent would be able to name the top three priorities. When *Inc.* then followed up with the employees, only 2 percent could accurately name their leader's priorities.[14] So, there's a good chance your team needs a few reminders about what is actually a priority, and what just appears to be so. Too many false priorities create ambiguity. And ambiguity is the enemy of clarity (not really, it's actually just the antonym of clarity, but the point is still the same).

Too many false priorities create ambiguity.

To make priorities clear in the military, many strategists specify *commander's intent*, a clear description of how the commander of the operation defines a successful mission. It's a literal description of what success looks like. It gives everyone a clear understanding of priorities. When chaos and confusion ensue, *commander's intent* also empowers subordinates to improvise and adapt, so long as they're still working together on achieving their commander's definition of success.

You may not need to issue battlefield commands, but helping the team set priorities will likewise empower them to coordinate their tasks and structure their calendars, ensuring that critical tasks aren't ignored while half the team spends their days clearing out their emails. In addition, when project pivots happen or when a new project is started, it's time to revisit priorities and make clear where each project lands in the rankings.

SET IF-THEN PLANS

Making priorities clear is crucial, but so is making a plan for keeping focused on those priorities. As the principle of commander's intent demonstrates, chaos happens. When clarity turns to confusion, many individuals and teams get stuck making decisions and taking actions. But psychology research reveals a hack for cutting through the confusion. Humans are quite good at remembering information when it's presented in "If x, then y" format.[15] ("If it's Monday at 9 a.m., then I'll send a huddle update to my team.") When people decide when, where, and under what conditions they will take action, they're more likely to fulfill their goals. Our brains begin to link certain behaviors to certain situations or triggers. Once "if" is detected, "then" goes to work.

Likewise in teams, If-Then plans can help coordinate team actions and foster accountability. ("If I get the outline from Sarah, then I'll go to work creating charts for the slide deck within the next day.") When you set If-Then plans, there is less room for conflicting interpretations around who is doing what and when. The actions, triggers, and time frame have been set out in advance. If-Then plans can even help teams anticipate unforeseen circumstances and coordinate a response without wasting time. Those types of If-Then plans are best set at the beginning of the project when considering possible derailers and deciding on the favored response. ("If the budget gets cut in Q2, then we'll reduce marketing expenses first.")

And if the syntax of "If x, then y" sounds clunky and lame to you, then don't worry. (See what I did there?) It does sound unnatural compared to how people tend to communicate. But that's a good thing. The abnormal phrasing makes the plan easier to remember, which makes it more likely people will actually take action.

Start small and choose one of the above to start implementing with your team. When everyone is clear on how it works, add another one into your routine. You'll watch the level of clarity quickly rise beyond your expectations. And when teams not only understand what to expect from others, but also truly understand others on their team, their performance will rise even higher.

And empathy is where we're headed in the next chapter.

CHAPTER TWO

EMPATHY

THE UCLA WOMEN'S GYMNASTICS PROGRAM IS ONE OF the best in the country. Former Head Coach Valerie Kondos Field was one of the best coaches in the world. Under her tenure as head coach, UCLA won seven national championships and nineteen conference titles. They're the only college gymnastics program to not only recruit Olympic gymnasts but also send alumni gymnasts to the Olympics—a feat done four separate times for two different national teams. Kondos Field was named gymnastics national coach of the year four times and PAC-12 coach of the year five times. She was inducted into the UCLA Hall of Fame in 2010 and was named PAC-12 coach of the century in 2016.

All of this is a little surprising, because Kondos Field is not a gymnast and never was.

"I have never done a cartwheel," she has said. "And I couldn't teach you how to do a proper cartwheel."[16]

In fact, she doesn't even like being called "Coach." Coming from the world of dance, she has always preferred her student athletes to address her as "Miss Val."

Kondos Field was trained as a ballerina and came to gymnastics as a choreographer. Actually, she tried to enter gymnastics coaching as a choreographer, but the gym she applied for only wanted to hire her because she could play piano. (At the time, gymnasts' floor routines were accompanied by live music, not recordings—something the Olympics should maybe reconsider. Imagine the gold medal rounds of Olympic competition with Billy Joel or Beyoncé playing on the sidelines.)

Her work with that gym gave her the confidence to apply for a choreographer job at UCLA, which offered her an assistant coach position and a scholarship to attend the university. In 1990, Kondos Field was offered the head coaching position, a surprise honor she accepted thinking it would be temporary. She told the athletic director she would take the job for a year while UCLA found a qualified head coach, one who actually knew something about gymnastics.

To try to grow into the role, Kondos Field did the only thing she could think to do: she imitated other successful gymnastic coaches. She became a tough-minded—and often unempathetic—dictatorial coach. At the time, the most renowned program was the Károlyi Ranch, run by Martha and Béla Károlyi, a notoriously brutal place where gymnasts were treated like interchangeable parts in a machine built for only one purpose: to win. The Ranch stripped gymnasts of their individuality, forcing them to wear their hair in a specific way, stand in a specific way, and respond in a specific way when spoken to. (The Ranch was also the setting for much of team doctor and convicted child molester Larry Nasser's abuse—something we'll unfortunately get to in a little bit).

Not knowing what else to do, Kondos Field started coaching her team the way the Károlyis did. The results were dismal. In her first year as head coach, UCLA finished dead last at the NCAA national championships. But being inundated in this world of athletics coaching had unlocked a competitive streak in Kondos Field. After the loss, she walked into the UCLA athletic director's office and asked for another year to prove herself.

In her second year as head coach, UCLA didn't even qualify for the national championships.

That's when her team got involved. They called a team meeting, and for two hours, they offered examples of how destructive Kondos Field's leadership style had been. They told her how she came off as arrogant and demeaning, adding that she was hurtful. "They explained to me that they wanted to be supported, not belittled," she reflected.[17]

It took a little bit of time to process such a hard hit; but, eventually, Kondos Field realized how much of a blessing it was—because it was the impetus for a change in coaching philosophy that was more authentic to who she was as a leader, a dancer, and (in her words) a Greek. "I realized that I needed to fortify our student athletes as whole human beings, not just athletes who won," she said.[18]

In other words, she realized that she and the team needed to learn how to empathize with each other's individuality—and, in doing so, support each other's development as people and eventual champions. That didn't mean no longer holding them accountable for their commitments to practice and compete, but it meant balancing that performance accountability with an understanding of the person.

Almost as soon as Kondos Field's new empathetic style emerged, the team started winning again, but on even bigger stages. They won the PAC-12 conference championship in 1993, 1995, and 1997—and in 1997, they won UCLA's first national title. Twelve more conference titles and six more national championships would follow. As they racked up wins, Kondos Field's empathic coaching would make an even bigger impact on her athletes and the sport.

In 2015, for instance, renowned gymnast Katelyn Ohashi decided to skip training for the Olympics and join UCLA instead. Katelyn struggled in her first season, eventually telling the team, "I just don't want to be great again." The brutal system of gymnastics she had been raised in had crushed her spirit and left her bitter. Ohashi rejected Kondos Field's coaching, at one point telling her "Everything you tell me to do, I do the exact opposite." But Kondos Field and the rest of the team were committed to understanding Ohashi and building trust with her. Kondos Field took the time to learn about Ohashi's whole self. While she continued to coach her in gymnastics, she committed to only doing so in the gym—and using their interactions outside the gym to discuss school, life, friends, hobbies, and everything else.

Through this genuine interest, Kondos Field found a way to interact with Ohashi and connect her back to what she loved about the sport. "It was so cool to see the process of Katelyn Ohashi blossoming before our eyes," Kondos Field said. For the rest of us outside of the team, that full bloom was on display in 2019, when Ohashi won a national title on floor exercise, and her energy and enthusiasm for the sport was captured on a video that eventually exceeded 200 million views online.

But the biggest test of Kondos Field's empathy and understanding happened when Kyla Ross walked into her office on an otherwise normal day. Ross was one of her gymnasts and a former Olympian. She was normally strictly business, but on that day she started by making small talk that wandered from college, to friends, to seemingly everything else. "My inner voice whispered to me that something was on her mind, and if I was still and gave her enough time, it would come out," said Kondos Field.[19]

It did.

Ross revealed that during her training for the Olympics, she had been sexually abused by Larry Nassar. It was the first time Ross had told anyone. The revelation put Kondos Field in a tight spot. She was obviously going to support Ross and help her in any way possible. But the national championships were just a few weeks away, and Kondos Field worried that what they were about to go through as a team would be a distraction. Still, she knew it was best for everyone to go through it together. The emotional well-being of her athletes was more important than winning titles.

With the support of Kondos Field, Kyla Ross and others went public with their allegations. Over time, Ross' decision to go public led to the discovery of more than 250 other victims Nassar had abused. And, eventually, it led to Nassar being convicted and jailed for life.

During all this, Kondos Field dedicated several team meetings to addressing the issue and creating a supportive environment for the team to both work through their trauma and train undistracted. As it would turn out, Kondos Field's worries about their performance at the national championships were unnecessary. UCLA won the women's national championship, and Ross won the individual championship in both floor and vault. After the event, Ross shared with Kondos Field her belief that confronting their trauma wasn't a liability—and that teaching the team to empathize and support each other was likely what led them to victory. "Miss Val, I literally felt myself walk taller as the season went on," Ross told Kondos Field. "And when I walked onto that championship floor, I felt invincible."[20]

Kondos Field's coaching career, like Chris Hadfield's space station mission, proves there's more to performance than just being clear on execution. There's a lot more, obviously, because there are two other parts to this book. Even when it comes to common understanding, teams perform better when they not only understand what other members are doing, but also understand each other as humans. In other words, they do better when clarity is paired with empathy.

Feeling What Empathic Teams Feel

It may seem strange to add empathy to the list of cultural elements high-performing teams share. Traditionally, we think of empathy as something experienced by family members or friends. We think of it as personal, and not business. But *The Godfather* was wrong. There's no such thing as "It's not personal, it's just business." As long as teams are composed of people, building empathy between those people will remain a vital component of great teams.

Empathy is when team members are able to understand the emotions and perspectives of others to an extent that they almost experience those same emotions. (Empathy and sympathy are often used interchangeably. But if you're wondering how to easily explain the difference between them, here goes: sympathy is when you feel sorry for another person, and empathy is when you feel what the other person is feeling.) In the case of teams, empathy means that team members understand each other so well they can put themselves into each other's frame of mind. They understand the strengths and weaknesses, fears, struggles, and opinions of others on their team.

And you can see why that understanding would help teamwork and collaboration. It's one thing to know what to expect from your teammates. It's another to be able to comprehend what they're thinking as they're working or as you're planning work. That's why teams that achieve empathy perform better. They understand each other so well that they reach a whole new level of clarity. Call it clarity of *person* instead of clarity of roles.

To put it another way, empathy is vital to a high-performing team because great teams play chess, not checkers.[21]

Great teams play chess, not checkers.

In checkers, every piece is the same and moves in the same way. In chess, each piece is different and moves in a different way; you can't ignore the uniqueness of each piece. In order to win, the entire team needs to plan and coordinate based on different movements.

In order to appreciate the different moves different team members can execute, you need to understand more than just their job description; you need to understand *them*. You need to understand that they have different communication preferences than you, and that they signal for help differently and offer suggestions differently. And when everyone on the team gives up trying to create uniformity and instead seeks to empathize with the team's diversity, that's when they start to truly perform to their potential.

Remember Anita Williams Wooley and the collective intelligence studies from earlier? Wooley and her team found that the largest predictor of whether a team could reach the level of empathy necessary for optimal performance was how much *social sensitivity* existed on the team.[22] In other words, how sensitive the team was to the differences of others was a significant determinant of their success. (Interestingly, the strongest correlation of team success was actually the number of women on the team. But the researchers argued that since women tend to be more socially sensitive than others, social sensitivity was the most likely explanation. Still, one easy shortcut to teamwide empathy might be to just add more women to the team.)[23]

When teams have strong social sensitivity, they're able to respond better to subtle cues from their teammates. The conversations and collaboration become more balanced, with everyone sharing their perspective and ideas equally instead of one or two members of the team dominating. "Social sensitivity has to do with how well group members perceive each other's emotions," said Christopher Chabris, another of the researchers on Williams' team.[24] In other words, it's all about how much empathy the team shared.

Building empathy on a team is a bit of a mysterious process. There are a few hacks and shortcuts to empathy (like adding more women), but often it requires time to develop—time spent doing nonwork activities. On teams, empathy usually forms by understanding the whole person, not just their job description. That requires talking about more than just work, and being willing to share more about themselves than just their work-related attributes. The simplest way to start this process is often to spend time together doing what looks unproductive, like sharing meals or simply hanging out together.

Perhaps that's why Kevin Kniffin and a team of researchers found that fire stations that share regular meals together perform significantly better.[25] Breaking bread is one of humanity's oldest rituals, and it's one of the earliest ways we learned to build connection and empathy with our fellow humans. The understanding about each other that develops at mealtime carries over to just about every other task a team is asked to perform. In fact, that particular study even found that in stations where shared meals weren't the norm, firefighters were often embarrassed to explain why. They knew intuitively that something was broken on the team, and they were embarrassed that their teammates weren't working to resolve the issue with such a simple fix.

Teams without empathy feel like those embarrassed firefighters. They fight for dominance in team discussions. They argue over who gets the choice roles on the team and the preferred tasks or assignments. They know that good teamwork is supposed to feel different than this, but since it doesn't, they turn to looking out for themselves first and pay little mind to who on the team gets left behind in the process.

But teams with empathy look out for each other. Teammates with empathy put the needs of the team and the perspective of others first when discussing ideas or planning projects. They don't finalize commitments until everyone understands the plan and agrees with it—and until the plan perfectly leverages the strengths and ideas of each member of the team. When that happens, teams perform at their best—because they don't just understand what to do; they also understand why they're doing it and the way they're doing it.

Enhancing Empathy

You might be embarrassed by the low level of empathy on your team, or it might be so high that your team feels like a family. Regardless of where you are, empathy is a process. At each stage of the process, there are ways to move further along and create even more empathy on the team. Here are a few tactics that have worked for some of the highest-performing, most empathetic teams:

WRITE TEAMMATE MANUALS

In the last chapter, we covered the value of team charters for setting clear norms about collaboration and communication, but what team charters can leave out are the personal preferences and behaviors of individual teammates. In a teammate manual, or a "user's manual for me," each person drafts a short document telling their teammates more about themselves and how they prefer to work. These manuals help the team understand why one person always seems overly optimistic while another seems skeptical, and why one person writes long, contemplative emails while another writes back, "Sounds good." This saves time and confusion and also helps reduce conflict, perhaps better than any over-priced personality test could. (People don't need to know what color or series of letters their teammate is; they need to know specifics, like why he always responds to questions with another question.)

One easy template to start with contains four simple statements:

- I am at my best when _____.
- I am at my worst when _____.
- You can count on me to _____.
- What I need from you is _____.

Send these questions out and ask the team to ponder them for a while before meeting to share answers. If you're the leader, establish trust by going first (more on that in Part Two). Allow time after each statement for questions and clarification, as people are trying to apply what has been shared to past experiences with that person. Just like team charters, the real value is not in the document, but in the drafting and sharing of it.

FIND FREE TIME

One of the most productive times for team collaboration is when the team does nothing at all. That sounds counterintuitive, but humans are social creatures, and socialization is how we learn about each other best. In times when people aren't talking about work, they're usually talking about themselves. They're describing past experiences, introducing their family, and sharing hobbies and interests that extend beyond their job description and training.

These moments of self-disclosure allow the whole team to understand the person better, and they allow individual teammates to find *uncommon commonalities*—things that those two have in common but that are uncommon to the rest of the team. These uncommon commonalities are how individuals build bonds and how coworkers turn into friends. A myriad of research suggests having friends at work and on a team makes people more productive, engaged, and resilient.

Some unstructured times happen naturally, like the moments before a meeting when some of the team is in the conference room or on the video call early. But other times may need to be created deliberately, like setting certain days to eat together or creating a calendar of paired "coffee chat" appointments between coworkers. These deliberate times might seem "fundatory" (mandatory fun that's not actually that fun), but that's likely because the team doesn't know that much about each other yet. As these times continue, and as the team grows closer and develops more empathy, they'll quickly turn into some of the most energizing times on a team's calendar.

BREAK THE ICE

Building on the uncommon commonalities tactic, using icebreakers to kick off or close meetings helps teams learn more about each other in a more direct and targeted way. And yes, if you're closing a meeting with an icebreaker, it's not technically an icebreaker—but that's the term everyone loves to hate. And also, yes, I know you've probably been in some awkward and cringey icebreakers in the past, but—interestingly enough—research suggests that icebreaker-type rituals can still improve employee's commitment even when participants feel a little awkward.[26] In fact, feeling awkward is the point. It's a small-scale vulnerability and disclosure that not only builds empathy, but also trust.

There's also a myriad of non-cringey, yet strikingly effective, icebreakers to choose from. A few honorable mentions include:

- *Triple H*: Ask each member of the team to share a hero, a highlight, and a hardship. You'll learn an awful lot about what someone values and how they see their strengths and weaknesses through just these three answers.

- *Energy Check*: To kick off a team-wide meeting, ask each member to rate their level of energy on a scale of one (dragging) to five (jumping). Then, ask what they or the team could do to help bring that energy up one level (unless, of course they answer "five"in which case you should ask them to stop jumping and sit down so others can share). You'll learn a lot about each other's work habits and work needs just from asking.

- *Defining Moment*: Have each member of the team share the story of a single moment in their life that helped shape who they are today. Don't get more specific than that. Let them define "shape" however they choose.

- *Three Snaps:* Before each meeting, ask a different team member to choose three photos from their phone's camera roll to share with the group. Have the team member tell the story behind why they chose those three snaps.

SHARE GRATITUDE

One of the simplest and most powerful ways to build empathy and connection with someone else is to show appreciation. So, it's not surprising that research suggests high-performing teams express significantly more gratitude to each other than other groups. In addition, increasing expressions of gratitude on a team also increases the openness to helping each other on future projects. The benefits of gratitude aren't just reserved for the receiver; they're also gotten by the giver. (Please forgive the grammar there in favor of some awesome alliteration.)

Taking the time to say "thank you" increases well-being and brain function and reduces impatience and other stressors that get in the way of empathizing with colleagues. Grateful people are more relaxed and more resilient, and they earn about seven percent more than their ungrateful colleagues.

Consider starting a few public displays of appreciation on your team. This could be a weekly ritual at the end of a meeting where each person says thanks to someone else (and pay attention: you want to make sure everyone receives at least one kudos). It could also be by creating a "Weekly Praise" email or communication channel where members share what they appreciated about each other in the past week. If you need an even smaller start, you could target just one person and pass around a symbol or token when they receive appreciation (having the token would also nominate them to share during the following week).

You may have found something on this list your team already does, or you may have found something you tried once and then stopped for some reason. In either case, keep going. These all work best when they stop being a random activity and start becoming a habit on your team. But if it seems like these activities are too cringey or too difficult for your team, it may be because teammates don't feel comfortable opening up yet. They don't feel safe enough, yet.

Luckily for them, safety is what we're focused on in Part Two.

PART TWO

PSYCHOLOGICAL SAFETY

Common Understanding

Psychological Safety

Trust + Respect

Prosocial Purpose

IN 2015, GOOGLE PEOPLE ANALYTICS TEAM LAUNCHED one of the most ambitious investigations into team performance ever.

And they found nothing.

At first.

Led by Julia Rozovsky, the analytics team embarked on a multi-year study to answer the question, "What makes a team effective at Google?" They called their investigation Project Aristotle, after the philosopher who coined the phrase "the whole is greater than the sum of its parts." They wanted to know what ingredients made a team's performance more than just the sum of its performers. More ambitiously, they wanted to discover it by studying real-world teams at one of the world's largest companies—not research participants in a lab.

Rozovsky's crew assembled data from 180 different teams across all departments, divisions, and countries at Google. The teams ranged in size from three to fifty individuals (although the median was just nine). The researchers conducted hundreds of interviews, gathered reams of performance data, and collected 250 different items from existing surveys given at Google. At first, they turned their attention to team composition—to the attributes of the people on their team. They looked at educational backgrounds, hobbies, friends, personality traits, and skill sets. They looked at the level of diversity on the team and how long the team had worked together. They considered overall tenure and how much of the team was located together (and even where they were located, just to see if some offices were better than others).

And they found nothing.

"We imagined it was going to be pretty algorithmic," recalled Rozovsky. "We thought it was probably going to be something like: take a few ex-consultants, pair them up with some rockstar engineers, get some extroverts. Voilà. You have a rockstar team." They thought once they had the data, a formula for perfect teamwork would emerge—a recipe for cooking up the ideal team. "We were just completely wrong."

No matter which way Rozovsky's team examined the data, they couldn't find a signal in all the noise. Undaunted, they went back to the existing research literature to see if there was another way to analyze everything they'd collected. There was. In looking back over the existing research and their data, they realized they had never considered something: norms.

Every team has norms. Traditions, behaviors, and unwritten rules govern how every team functions. Teams develop habits of how they interact. But Rozovsky and her team realized they hadn't yet considered group norms.

So, they went back to their interviews and data on performance and found norms everywhere. Some teams took turns ensuring everyone was allowed to speak; others interrupted each other constantly. Some teams took time before and after meetings to interact socially; others showed up and jumped right into the agenda. Some teams disagreed wildly but stayed civil; others broke down into arguing at even a hint of disagreement. These norms seemed to affect the performance of the team.

"We found something we didn't expect," said Rozovsky. "How a team works together was so much more important than who specifically was on the team."[27]

If you've gotten this far in this book, that insight shouldn't come as a surprise. But to the People Analytics team in 2015, it was a massive shock. It meant their approach to performance couldn't just focus on finding the most talented people and partnering them up together. It had to center on teaching teams the best ways to work together.

Specifically, the researchers found five common themes that all high-performing teams shared. And chief among those five was a feeling of psychological safety. (If you're wondering, you can read about all five in this book's endnotes—but let's just say the other four correlate nicely with Common Understanding and Prosocial Purpose.)

Psychological safety refers to how safe the team felt to express themselves and take risks. Teams with psychological safety have members who can be vulnerable and authentic with each other. They ask questions or offer ideas that may seem odd but can lead the team's thinking in new directions.

"Psychological safety actually drives performance," said Rozovsky. "And the reason that this happens is that when team members feel safe, they're much more likely to ask for help. They're much more likely to admit a mistake. They're much more likely to try new roles and responsibilities, knowing that their team has their back—and, by doing this, they learn. And by learning, they become more effective."[28]

Few teams take that approach to making mistakes, because few teams have psychological safety. Of the three elements of the best teams ever, psychological safety may be the hardest to build. It's certainly the rarest, because it involves totally rethinking how leaders and teams respond to divergent ideas and mistakes. In fact, how well a team handles mistakes is kind of how psychological safety was discovered in the first place.

Finding Safety

The term "psychological safety" was coined all the way back in 1965, in a discussion between leadership thinkers Edgar Schein and Warren Bennis about how to help people become open to change. But it wasn't until thirty years later that we had an understanding of just how important psychological safety was to teams. Even then, we almost missed it.

In the late 1990s, Harvard Business School professor Amy Edmondson was studying the interplay between leadership and teams of nurses at a Boston hospital. She originally set out to determine the difference between the best and worst charge nurses—the nurses in charge of a particular floor or department of the hospital. "I didn't set out to study psychological safety," Edmondson admitted. "But rather to study teamwork and its relationship to mistakes."[29]

But after using team evaluations of leaders to separate the good from the bad, she noticed something peculiar. The teams with better quality leaders had a higher rate of documented errors than teams with poor leaders. In that one metric of performance, good leadership seemed to be making the team worse. It was puzzling, and it was tempting to dismiss it as an outlier and move on.

But later, during interviews with individual nurses, Edmondson's research team stumbled upon the answer. It wasn't the mistakes; it was the documentation. The better charge nurses created an environment where people felt safe to admit errors, and so more errors got documented. On teams with poor charge nurses, the individual nurses were more likely to hide their mistakes for fear of punishment. Besides being massively unethical, hiding mistakes meant preventing the rest of the team from learning.

Edmondson used the term coined by Schein and Bennis and argued that good leaders created psychological safety that allowed people to admit failures and allowed the whole team to learn from those failures. That wasn't the only effect psychological safety seemed to have on team performance.

Psychological safety encourages team members to speak up when they disagree; as a result, more diverse viewpoints are shared. Psychological safety reduces failures, because when people feel that they can speak freely, they're more likely to intervene before a team makes a mistake. And, because fear triggers the amygdala with the "fight or flight" response that can shut down working memory and information processing, psychological safety (which reduces or eliminates fear) even enhances clarity of thought.

Psychological safety increases how quickly a team learns and adapts, and that increases team performance. But psychological safety doesn't mean each team member isn't accountable for performance. Individuals still know they are responsible for delivering on their promises to the team, and that there are consequences for shirking that responsibility. It means team members acknowledge that failures happen—especially failures triggered by circumstances outside their control—and that when they do happen, the team will focus on extracting lessons from failure, not executing punishment.

It's also important not to mistake psychological safety for groupthink. Teams aren't psychologically safe because everyone is close friends and they never disagree. It doesn't mean there are no problems. And it certainly doesn't mean they all think alike. Instead, psychological safety happens when teams develop norms of behavior that allow productive, task-focused conflict without devolving into personal squabbles. It's not a team where everyone tries to think alike. Instead, Edmondson describes it as "a team climate characterized by interpersonal trust and mutual respect in which people are comfortable being themselves."[30]

We'll come back to those ideas of interpersonal trust and mutual respect, but it's worth pausing to consider first what he meant by "being themselves."

Most leaders know that to build a high-performing team, they'll need to build a diverse team. In fact, it's one of the assumptions Julia Rozovsky and the People Analytics team at Google started with. But without psychological safety, diversity can actually be a liability, not an asset.[31] In a study conducted by Edmondson and Henrik Bresman from INSEAD business school, diversity had a slightly negative effect on performance on average.[32] However, the correlation flipped in teams with high psychological safety. For teams with really low psychological safety, diversity became an even greater detractor.

Diversity without safety
is anarchy.

Diverse teams experience more conflict, higher turnover, and more difficulty with communication and collaboration. Diversity without safety is anarchy. However, when teams develop a sense that the team is safe for interpersonal risk-taking, and when they understand that conflicting ideas are one of the most valuable forms of collaboration, they learn faster and perform better. This doesn't mean leaders shouldn't pursue diversity. It just means that without that safety, embracing diversity for its own sake is an invitation for conflict.

So, how do we create a climate of psychological safety on a team? Well, the answer is found inside Edmondson's description. It's a climate characterized by interpersonal trust and mutual respect.

Trust and Respect.

These qualities may seem similar. But they have their differences. The interplay between them is what builds psychological safety. Trust is how much we feel we can share our authentic selves with others. Respect is how much we feel they accept that self. If I trust you, then I will share honestly with you. If you respect me, then you will value what I've shared.

Given this interplay, it's worth taking the time to consider separately the roles of trust and respect on teams, as well as how to create them. And now is a good time for it, because trust and respect are our focus for the rest of Part Two. We'll review the research on what trust really means and how to create respectful, civil behaviors on a team. We'll also look closely at two organizations that weren't exactly paragons of trust and respect—at least not until a change of leadership led to a change in culture.

CHAPTER THREE

TRUST

ON HIS FIRST DAY AS THE NEW CEO OF FORD MOTOR
Company, a reporter asked Alan Mulally, "What kind of car do you drive?"

He replied without hesitation, "A Lexus. It's the finest car in the world."

It's worth reading that again.

On the very first day—the very first press conference—of his leadership at Ford Motor Company, Alan Mulally told a reporter he drove a competitor's car. Not only that, but he also called it the finest car in the world.

The reporter followed up, asking, "So what does that mean to us at Ford, if you think Lexus is the finest car on the market?"

Alan then replied, "That is why I'm excited to join Ford to help create the best cars and trucks in the world."[33] That's how committed he was to speaking the truth—and building a culture at Ford where everyone trusted each other enough to speak the truth as they saw it.

Bill Ford, the great-grandson of founder Henry Ford, asked Mulally to lead an effort to turn around the struggling automaker. Mulally had previously led a turnaround effort at Boeing when the terrorist attacks of September 11, 2001, led to nearly every airline canceling their production orders at Boeing. Though he lacked any experience in the auto industry, Ford believed his experience saving one of the world's largest aerospace manufacturers would transfer. Mulally knew that all the automobile experience in the world wouldn't matter if he couldn't significantly improve Ford's company culture, which is why he brought with him the same "Working Together" leadership and management system he'd refined and used at Boeing to transform their culture from one of command and control to one marked by cohesiveness and collaboration.[34]

At the time Mulally started, Ford had significant viability issues. And that's the most polite way to describe it. The business model was outdated and flawed. Ford's response to increased globalization was to create as many different cars as possible, with as many different production processes as required—leaving little ability to leverage economies of scale. But more than the business model, the culture was failing at Ford. Senior leaders, and even middle managers, were constantly vying for position—placing their desire for career advancement over corporate success.

The year Mulally took over, the company's best estimates were that it was going to lose $17 billion. Mulally's plan to turn Ford around involved significantly improving the culture. In order to confront the brutal facts and chart a path out of danger, Mulally knew he needed his leadership team to trust each other and trust him. He knew the team needed to come together around a compelling vision, comprehensive strategy, and relentless implementation to create value for all the stakeholders and the greater good. He knew he needed them to speak freely if they were going to understand just how dire the situation was—and if they were ever going to find a fix. In the first few weeks of his tenure, he announced he was changing how the leadership team would work together. Every Thursday morning at 7 a.m., the entire leadership team would gather in the Thunderbird conference room at company headquarters. Executives would review 320 slides that described the plan and the status of the plan—everything from production schedule to revenue streams. Senior leaders would be responsible for knowing their contribution to the plan and its status. They would take turns sharing what progress they'd made, what their focus was, and what was blocking their progress.

If that sounds a bit like a huddle from Chapter One, there's a good reason for it: it's exactly like a huddle.

"We're not going to be able to manage a secret," Mulally once told them when he explained the rationale for these meetings. "The idea is that we can share what the situation is and help each other."[35] To help them help each other, Mulally asked the team to color-code their slides. Green slides would mean everything was on track. Yellow slides would mean there were potential problems but they had a solution, and red slides would mean there was a problem without a current solution identified. To Alan, yellow and red slides were "gems"—they pointed to the issues that needed fixing in order to turn around the company. But to the leadership team he inherited, they meant something totally different: career risk.

The first huddle—Mulally called it a Business Plan Review—was held on September 28, 2006. And every single slide was green.

For weeks, Mulally sat at the boardroom table of a company bleeding money and listened as its senior executives each told him their departments weren't the problem. The situation was worse than Mulally had thought. The executives were afraid. They felt in danger. They felt it must have been a trick. The idea of being open and transparent with the very same people who were trying to sabotage their job was too foreign for them.

Several weeks in, Mulally stopped the meeting. It was another week of green on green on green, despite the company being in the red. "We're going to lose seventeen billion dollars this year, and all the charts are green," he shared. "Do you think there's *anything* that's not going well? Maybe even just one little thing?"[36]

Mark Fields, the Executive Vice President of Ford Europe, whom many assumed would be Bill Ford's successor before Mulally showed up, was starting to wonder if Mulally was serious about all that trust and candor talk. Fields oversaw the production of the Ford Edge, which was scheduled to launch in just a few weeks. But production problems at Ford's Oakville, Ontario facility had left thousands of cars unable to pass quality inspection and forced the plant to stop production. The day before, as Fields was running his own business plan review with his team, they discussed the issue. But his team had left the update slides green. "This sounds like one of those red situations Alan is always talking about," Fields told his team. They agreed, but also warned Fields that being the first to present a problem to the new CEO might mean the end of Field's career at Ford. Still, production had stopped—that fact couldn't be denied.

Without any other ideas, Fields did the right thing and created the transparency that Mulally was encouraging so they could work on the "gems" together, and he changed the color of his launch slide to red.

"On the Edge launch, we're red," Fields said at the following week's meeting. "We're holding the launch and finding a solution."[37]

The room went silent. The other executives stared at Fields, many of them thinking they'd better get a good look at him now, as they would likely never see him again. They wondered if two security officers were about to come barging through the door to escort Mark away.

Then, someone broke the silence. It was Mulally. He started clapping.

"Mark, that is great visibility," Mulally said, smiling. He asked the team for their ideas to help Mark with the situation.

Despite being stunned, several executives raised their hands and offered ideas and even members of their teams to help Fields find and fix the problem. Mulally knew they were now making progress on working together as a team. But progress was slow.

The next week, the executives filed into the same conference room, with the same set of green slides. But when they walked in, they saw something they *really* weren't expecting. Fields hadn't been fired or demoted. He still had his job. In fact, he was already in the conference room when the others arrived, sitting right next to Alan Mulally. Mulally would later call that morning the defining moment in Ford's turnaround. When the other executives saw that Fields hadn't been punished but actually commended for his transparency, they were finally willing to trust Mulally—and each other.

The following week, the slides were colored like a rainbow and started to accurately capture the company's situation. The team was ready to trust—and ready to confront the extremely serious situation they were truly in. Little by little, aided by that new level of trust, they started to assemble a turnaround plan by working together. And, as the executives started holding their own business plan reviews with the teams they led, they found that Mulally's "Working Together" plan started working really well.

Trust and transparency flowed from the Thunderbird room throughout the extended enterprise and flowed back with timely updates and changes to the culture. Eventually, that culture change also positively changed the performance of all stakeholders. Ford was on its way to profitability—so much so that when the financial crisis of 2008 and 2009 hit, Ford turned down the bailout money offered by the US government. They were the only one of the "big three" American automakers to do so, because they were the only one of the three that had successfully improved their culture to create near-term and long-term value for stakeholders and the greater good.

By 2010 they delivered a full-year profit of $6.6 billion. Mulally's focus on trust and transparency had taken them from losing $17 billion to profiting nearly $7 billion. When he retired, Mulally had built the most trusted and high-performing teams Ford had ever seen and the number one automobile brand in the United States.

Trust Begets Trust

Building trust made a huge difference at Ford. It makes the same difference on every other team. Trust on a team acts as a social lubricant. When teams trust each other, social frictions are reduced, which makes it easier to collaborate and easier to share one's true perspective on any challenges. Without trust, people hold back their brilliance. They don't share their unique perspectives and insights. With trust, those hesitations disappear, and ideas and information flow freely. Teams with trust explore more possibilities than teams without trust. And teams with trust willingly put forward their ideas and aid to others on the team. Trust even makes working on a team more enjoyable, even when it's working on a team trying to stop losing $17 billion a year.

In a 2011 study of nearly two hundred teams from the United States and Hong Kong, researchers found that when trust went up, so did performance. Interestingly, researchers John Schaubroeck, Simon Lam, and Ann Chunyan Peng distinguished two types of trust: cognition-based trust and affect-based trust. Cognition-based trust was how well the team trusted that the others were competent and reliable (we might call that part of common understanding). Affect-based trust was how much the team genuinely cared for each other as people and wanted to see each other succeed (this is the type of trust that is part of psychological safety). Both forms of trust were important, but affect-based trust increased performance significantly more than cognition-based trust alone.[38]

Further research helps explain why. According to Paul Zak, one of the foremost researchers on trust, when team members experience high levels of trust, they are 74 percent less stressed, and they report 106 percent more energy at work. They are 76 percent more likely to be engaged, and they are 50 percent more productive. The benefits of a trusting culture are even experienced outside of work. Members of high-trust teams experience 40 percent less burnout and 29 percent more satisfaction with their overall lives.[39]

So how does trust build between individuals and on a team? Zak's research points to something surprising about trust. In Zak's words, "trust begets trust." Despite the common sayings, trust isn't given nor earned. Trust is reciprocated. Zak found that humans "feel" trust because of the brain chemical oxytocin.[40] There are a lot of biological triggers for oxytocin. Physical touch, intimacy, and even childbirth stimulate the body's release of oxytocin. But one previously unknown oxytocin trigger was the feeling of being trusted by someone else.

For the last twenty years, Zak and his team have run a variety of different experiments, all with the same theme. Participants are recruited and placed in a condition where they will have to decide whether to trust a stranger. The more that participants feel trusted by the stranger, the more they will trust the stranger in turn.

One variation of the study asks a pair of participants to play a simple game where they trade money. Player One is given a sum of money and told that any money he sends to Player Two will be tripled—and that Player Two can send money back, but has no obligation to return any of the newly tripled sum. Logically, this game should fall apart. Player One should take the money and run. And if Player One is dumb enough to send money to Player Two, she should take the money and run three times as fast.

But that's not what happens. Often, a Player One chooses to trust a Player Two. And when he does, she responds in kind by sending money back. After the exchange, all participants are taken to have their blood drawn. That's where the study gets even more surprising—because Zak and his team found that the participants with more oxytocin circulating in their blood had acted more trusting toward each other when playing the game. And yes, the participants also had blood drawn before playing the game, so the researchers could tell if the game itself had triggered oxytocin release. In a later variation of the game, Zak's team even artificially administered synthetic oxytocin (via a nasal spray, if you want to get really nerdy about it). The people who received oxytocin showed much more trust in strangers than those who'd received a placebo. In fact, use of the oxytocin spray actually doubled the number of people who were willing to send *all* of their money to the other player.

"Oxytocin rises when someone trusts you and it facilitates trustworthiness," Zak explained.[41]

This is why Mark Fields' admission of failure was so critical to turning around Ford. And it's why Alan Mulally was willing to wait weeks for the first executive to admit to a problem. He knew there was no way such a toxic culture would believe his mere words about being trusting teammates. They had to see an act of vulnerability by someone—they had to feel trusted—before it triggered trust in their own brain.

Training Trust

In teams, this means that trust can't just be instantly created (unless you've got a case of oxytocin nasal spray on hand). Instead, trust develops out of the experience of working together. It happens most often when someone is willing to signal vulnerability first. Often, this person is the leader. That initial signal begins a virtuous cycle of trust.

There are a variety of ways to start the cycle. What follows are a few of the more evidence-based trust-building experiences leaders and teams can share:

SIGNAL VULNERABILITY

As we saw from Paul Zak's research, trust begets trust. When in doubt, it's the team leader who should go first. Signaling vulnerability and fallibility is not a sign of weakness. It's a sign of trust. Leaders take responsibility for their mistakes and are willing to admit when they don't know the answer. This may come as a surprise, since many organizational cultures reward the most confident, least vulnerable people with new management positions. But continuing to appear immaculate doesn't inspire the team to trust them. Instead, it inspires them to pretend to be perfect themselves—and diminishes trust over time. Signaling vulnerability also signals interdependence. We can't do our jobs alone. If we could, we wouldn't be on the team. Leaders who preach team effort, but don't own up to their own shortcomings (and hence team learning opportunities), pretty quickly lose their credibility and the trust of their team.

And signaling vulnerability creates a powerful moment where the rest of the team can begin to share its weaknesses, as well. Doing so makes the whole team stronger through trust. If there hasn't been a moment like this on your team in a while, then it's a good time for another one. You may have to be *selectively vulnerable*, meaning lead with something that's a clear weakness, but not a deal-breaker for your leadership or too much personal information for your team to handle. It may be enough to just acknowledge your emotions at a certain point in a meeting ("Sorry, I was overwhelmed there for a second.") And a simple "I don't know" can work magic when it comes to making people feel trusted. If you're constantly signaling vulnerability, but no one is returning the message, then you may need to ask a member of the team whom you personally trust to do so for you. Like Alan Mulally and Ford, sometimes teams need to see a teammate admit their mistakes and not be punished in order to believe their leader is serious.

CELEBRATE FAILURES

One way to create that "Mark Fields" moment on a team is to celebrate failures. This doesn't mean teams throw a party every time they lose, but it also doesn't mean that every loss immediately triggers a round of "shift the blame" or that they forbid each other from talking about "the project which shall not be named." Failures are inevitable, and often for reasons outside of a team's control. Clients change their minds. Budgets get cut. Global pandemics disrupt the supply chain and force everyone to look at each other on video calls. In order to build trust on a team, the team has to be comfortable with the idea that they will fail—and learn from failure.

So, taking the time to celebrate what the painful experience taught the team can be a worthwhile exercise. This happens in a number of ways. You could draft a "failure résumé" for yourself and encourage teammates to do the same, listing every job or project that didn't turn out as hoped. As a team, you could create a "failure wall" with pictures or quotes from projects that blew up or clients you didn't win. Sara Blakely, founder of Spanx, throws regular "Oops Meetings," where she admits her own mistakes and encourages the team to do the same.[42] One pharmaceutical company went so far as to create "Drug Wakes" to gather researchers together around a promised but failed compound. The team said their goodbyes and expressed gratitude for the lessons working on that aborted drug taught them.[43] These types of celebrations not only focus the team on lessons learned, but they also encourage future risk-taking and keep teams motivated even when those chances of failure are high.

For nearly two decades now, I've trained and coached others in the martial art of Brazilian Jiujitsu. We tell every new student the same thing before they head onto the mat for their first competition: you can't really lose.

You win. Or you learn.

SHARE PRIVILEGED INTEL

Knowledge is power. And one way to measure the amount of trust on a team is by looking at how much information people keep to themselves (low trust) or share openly (high trust). Team leaders signal that they trust their team when they share privileged information with them. This could be sharing the team's finances, or the budget handed down from higher-ups, so that the whole team knows what they're working with and where the priorities are. It could also be competitive data or customer trends that are widely available but rarely shared. It could even be passing on information shared by another team. In a meeting, any time someone says, "This stays here," the atmosphere changes. People know whatever follows is spoken in trust, and they respond to that feeling of being trusted with trustworthy behavior. (Most of the time. If someone has proven themselves untrustworthy, then this may not be the best intervention.) People begin to feel like they're in the inner circle when they receive more information about their work or the environment than they usually do. They start to see how their work fits into the larger organizational whole (more on why that matters so much in a few chapters).

One more caveat here. If you received information with the preface "this stays here," then you're not building trust by sharing it somewhere else. Honor the trust that was shown to you and find something else you can share to build trust with your own team.

CREATE RITUALS

This may not seem directly related to trust, but rituals can be a powerful way to create the sense of belonging it takes for people to start being vulnerable. Humans are social creatures. We "form" groups, tribes, and whole cultures. In every grouping of people, rituals develop as a method to signal membership. Rituals emphasize what's important about our interactions with each other and can be small, like special handshakes, or large, like promotion ceremonies. They can even be odd-sounding at first, like "failure wakes." But in each case, once people understand the ritual and the meaning behind it, they feel like insiders, and they feel more ready to trust other insiders.

Research even suggests that groups who perform rituals before engaging in tasks outperform their less-ritualistic competitors.[44] Even strange-looking rituals like a series of claps or patterns of waving arms were enough to increase performance on the team. The novelty of the ritual didn't matter; what mattered was that team members went through a shared experience—and, in doing so, they identified that much more with the team.

HOLD AFTER-ACTION REVIEWS

One ritual that works incredibly well for both performance and trust is the after-action review—although, unlike clapping or waving, this is a more serious ritual done after the action (hence the name). Originally a military ritual, after-action reviews work well because they force the team to discuss strengths and weaknesses and to dissect past failures (and even successes) for lessons.

Just after the team finishes a project, or during an important milestone, gather them together and ask a few questions:

- What was our intended result?
- What was the actual result?
- Why were they different?
- What will we do the same next time?
- What will we do differently next time?

The purpose of the meeting is not to find someone to blame, or someone to give all the credit to. The goal is to extract lessons from the project about where the team is strong and where they need improvement. When people are open and honest about their weaknesses and contributions to failure, celebrate the vulnerability they just signaled.

If you'd like to make after-action reviews a more frequent, less formal ritual for yourself or your team, consider a short, two-question version that can be done after every meeting or workday:

- What worked well?
- What would be even better?

Research has shown that even small amounts of time spent reflecting each day can add up to greater performance than just using the same time to simply dive into the next project.[45] Making reflection a ritual is how you make sure your team never loses. They win. Or they learn.

If you're starting with a brand-new team, or a team coming off a violation of trust, then building trust will take time. The virtuous cycle of trust is a flywheel—initially difficult to get moving, but picking up speed with every vulnerable moment and every lesson learned. But vulnerability and sharing information isn't enough. If candid moments are not met with respect, the flywheel might grind to a halt. So, in the next chapter, we'll cover what needs to happen after someone acts on trust.

CHAPTER FOUR

RESPECT

LOTS OF LEADERS CLAIM THEY LISTEN TO THEIR WHOLE team. Few actually do. Sure, they analyze employee surveys and read reports that get filtered up the hierarchy. Leaders want their people to trust them enough to share their input, but they fail to demonstrate enough respect to truly build that level of trust. A few take the time to get out onto the frontlines and show just how much they respect the input of everyone on the team, and amazing things happen when they do.

One of those remarkable leaders is Maggie Wilderotter. Wilderotter has a lot of achievements on her résumé. She has been the CEO of multiple companies. She has served as a director on the boards of over thirty companies. She's even one-half of the only pair of sisters ever to be CEOs of Fortune 500 companies at the same time.[46]

But her most amazing achievement is the one that's not about her at all. It's about the way she built a culture of trust and respect in a deeply divided organization.

Wilderotter was hired away from Microsoft to lead Frontier Communications' turnaround in 2004. At the time, Frontier (then called Citizens) was a communications company formed from the breakup of AT&T. It was profitable but operated mostly in low- or no-growth areas.[47] As a result, a potential sale of the company had fallen through, and the board decided they needed outside help if they were going to grow the company or attract a new buyer who would actually go through with the purchase.

Instead of asserting a bold new vision or copying the turnaround strategy of another company, Wilderotter decided that people inside Frontier already knew what to do. Senior leaders just needed to listen.

When Wilderotter joined the company, she quickly found she was actually leading two companies—with two very different ideas about what worked and what didn't. Like a lot of similar companies, Frontier had a class divide between white-collar employees at its headquarters in Norwalk, Connecticut, and the blue-collar workers who actually worked in the company's fifteen thousand markets across suburban and rural America. The gap between the two groups at this company was the widest she'd ever seen.

Despite interacting with customers directly, and despite their intricate knowledge of operations, employees in the field were almost never considered when making decisions. Executives rarely flew out to meet with regional leaders or frontline team members, even though Frontier owned a corporate jet with its own hangar and six pilots. They didn't respect the perspective of the people who actually knew how Frontier worked on a daily basis.

Wilderotter saw this immediately, even though the home office executives tried to steal her gaze. "Everybody wanted to show me the org chart, to make sure I understood the pecking order," Wilderotter recalled. "I didn't even look at it, because I believe that work gets done through the go-to people. They may not have titles and positions, but they're the ones who get the work done."[48]

Wilderotter needed their insight. But first, she needed to prove she respected it.

Within six months of Wilderotter's arrival, the company conducted an employee survey—its first survey ever. Wilderotter embarked on a listening tour, traveling to some of the most remote markets in the country to understand how things really worked and who her "go-to" people really were. And she did most of that travel flying commercial because she sold the private jet.

In order to move the company from two opposing groups to one unified team, Wilderotter was committed to giving workers a voice. She took the side of those on the front lines in nearly every dispute. It wasn't about who was right in some minor dispute about an irrelevant detail; it was about the principle of respecting everyone's voice. She fired the home office executives who weren't willing to listen as hard as she was. She gave frontline employees the first raise they'd seen in five years, a move that was less about the money than about demonstrating how much she respected their work and their opinions.

Wilderotter even dove into one of the most contentious issues in Frontier's industry: labor relations.

Before Wilderotter joined, the company outsourced negotiations with the union to its army of lawyers— sending the message, unintentionally or not, that such considerations were beneath leadership. Wilderotter hadn't worked with a union before but found the whole concept of general managers being uninvolved baffling.

"The union members worked for the general managers in Connecticut," she said. "So I said the general managers should negotiate directly with their employees." Wilderotter forced general managers to hear their employees' concerns and respect their needs and ideas. As a result, those needs got addressed. The union employees, those actually installing phones and fixing service issues, were added to the company's profit sharing and stock options. These changes helped align the focus of white-collar and blue-collar employees alike. "Having shared goals and targets helped make us all one company, helped build this circle of trust."[49]

During Wilderotter's eleven years of running Frontier, the company moved from a sleepy "Baby Bell" with revenues of $1 billion a year to a nationwide broadband provider with more than $10 billion in revenue generated from operations in twenty-nine states.[50] Perhaps more impressively, the union never went on strike once with her in the top leadership role.

Wilderotter proved that respecting employees makes them more likely to trust their leaders and feel psychologically safe. That, in turn, makes them more likely to contribute their ideas. And their ideas often reveal previously unknown ways to improve the company.

Respect Starts at the Top

Wilderotter's turnaround at Frontier serves as a compelling example of why psychological safety is more than just trust. Unless leaders and team members demonstrate they respect the differing opinions of others, whatever trust is built quickly dissipates. Without respect, the virtuous cycle of trust we're seeking to build can quickly reverse course to become a vicious cycle of distrust.

And, sadly, respect in most workplaces is as absent as it was at Frontier when Maggie Wilderotter took over leadership. At a minimum, feeling respected is a rare occurrence for most workers. In 2014, researchers Christine Porath and Tony Schwartz surveyed more than twenty thousand employees around the world and found that 54 percent said that they didn't get regular respect from their leader.[51] And getting respect, or not getting it, made a huge difference in their engagement and performance.

Employees who got respect from their leaders reported 89 percent higher satisfaction with work and 56 percent better well-being and overall health. They also reported 92 percent greater focus. Overall, they were 55 percent more engaged.[52] In fact, in their survey, Porath and Schwartz found that no other leadership behavior had a bigger effect on employee outcomes than respect toward subordinates.[53]

In a different study, Porath focused on those who had experienced incivility or disrespect at work. Among people who'd been disrespected at work, 48 percent said they intentionally decreased their effort, and 38 percent said they intentionally decreased the quality of their work. A full 66 percent said their performance declined, and 78 percent said their commitment to their employer declined.[54]

It's no wonder, then, that showing respect often affects long-term career performance, as well. Porath's research has shown that people who model respect tend to move toward central positions in the informal network of an organization. They become the connectors of ideas and people. They go out of their way to share information and provide help to others. As a result, they're more likely to be seen as leaders. By contrast, disrespectful behavior can limit career growth in organizations. High performers who receive too much disrespect from too many people in an organization become thirteen times more likely to quit than low or average performers in the same disrespectful culture—which has a massively negative effect on organizational performance.

And the same positive benefits of respect are found in the way teammates treat each other. In a study of incivility and respect in medical settings, twenty-four Israeli medical teams from four neonatal intensive care units participated in training to improve the quality of care for premature infants. As part of the training, teams simulated how they'd respond to an infant whose health was crashing due to an intestinal illness. The teams not only had to discover the intestinal cause, but they also had to treat it quickly. To compare the teams, all teams were told they'd be observed by an expert from the United States—but some teams were given a neutral greeting from the "expert," and others received an insulting message about the poor quality of Israeli medical care. They were flat-out disrespected, and you could see it in the performance. Disrespected teams didn't collaborate as well. They were less likely to diagnose the problem, and therefore less likely to begin the right procedure. The main cause was that teams who felt respected shared information more readily and were more willing to help, whereas disrespected teams stopped seeking help altogether.[55]

Teams without a sense of mutual respect shut down fast. They're less likely to seek feedback, and much less likely to receive it. They don't experiment as often. They rarely discuss errors or mistakes. Individual team members are less likely to speak up about problems they see—making total team failure more likely. In a medical setting, disrespect can be deadly. But it's also devastating to teams in all industries doing all types of work.

Respectful teams experience greater performance because members of those teams are more willing to go above and beyond. They're more likely to help teammates even when it's not required, and they're more likely to offer feedback that gets heard. Perhaps even more importantly, respected teammates are more likely to listen to what others say and actually consider differing opinions instead of fighting for their own. This means respectful teams are more creative and innovative, as well.

Respect is a learned behavior.

And it starts at the top. Across a series of surveys, Porath has often asked why disrespect happens on teams. (If you feel like you're reading Christine Porath's name often, it's because most of the research on respect comes from her lab. You could say she's the world's most respected researcher on respect.) The number-one reason Porath has been given for disrespectful behavior is that people are just modeling behavior seen elsewhere in the organization. Absent a role model of respectful behavior, team members copy the rude behavior of the disrespectful leaders in their organization.

In other words, respect is a learned behavior.

Leaders, in particular, teach teams how to behave respectfully toward each other. That's the reason Maggie Wilderotter spent so much time listening to the workers on the front lines and siding with them during debates with home office personnel. She was modeling the respectful behavior she wanted all executives to mimic. (And she was firing those who couldn't copy it well enough).

Because respect is contagious, many of the teams inside Frontier began acting more respectfully towards each other and other teams. In turn, that changed the culture at Frontier into one where people were more willing to share information and listen. While Frontier acquired several other communications businesses during Wilderotter's tenure, the fact that they acquired a sense of mutual respect contributed to their growth far more.

Give a Little Respect

In the last chapter, I outlined several activities and behaviors that can build trust on a team. But as we've seen, those activities alone won't create a climate of psychological safety on your team. You may succeed in building enough trust to allow people to share their authentic selves. But if that sharing isn't met with respect and acceptance, trust will have a half-life as short as oxytocin (three minutes, in case you were wondering). So, here are a few ways to pair respect on a team with the trust you're already working to build:

MODEL ACTIVE LISTENING

The easiest way to signal disrespect to someone is to make them feel ignored. The reverse is true, as well—making people feel listened to and truly heard is one of the simplest ways to signal that you respect what they have to say. Great team cultures are marked by how well they listen to each other and take turns speaking so everyone feels heard. But our natural tendency as humans can make it difficult to show others we're listening. We want to be helpful. When people come to us with problems, we want to jump in right away. For team leaders, this tendency is even stronger. People are supposed to come to us for help, right? So, we start helping...which means we start talking...which means we stop listening.

One simple trick for ensuring you listen longer and help others feel more heard is to get used to saying, "Tell me more." When someone says something that triggers a thought in your head, and you feel your mouth starting to open so your brilliant advice can greet the world—stop. Instead of whatever you were going to say, just say "Tell me more." If you want to take active listening even further, consider a useful acronym from communication expert Julian Treasure: RASA.[56] When someone else is speaking, *Receive* their ideas by paying attention to them as they speak. *Appreciate* what they are saying by nodding or giving confirming feedback. *Summarize* what the other person said when they're finished. Then *Ask* them questions to explore their ideas further. Since respect is a learned behavior, as you model active listening, your team will follow your example—and more members of your team will feel heard and respected.

RECOGNIZE, AND SHARE CREDIT

Leadership thinker Warren Bennis once noted that good leaders shine under the spotlight, but great leaders help others shine. Teams who share credit and take the time to recognize each other are teams where members feel more respected and more trusted. But teams who fight for credit when a project is finished (or fight over blame when it fails) diminish what little respect they had before. Great team leaders look for as many ways as possible to share credit with their team, even if they desire most of the credit for themselves. The act of sharing can be as simple as taking the time to appreciate each team member's strengths, or as elaborate as shouting those praises throughout the company. When team members know what you appreciate about them, they know you respect their abilities and ideas.

In addition, find small wins that can be celebrated more often, creating more opportunities to recognize others. Small wins have a big impact on individual and team motivation—and that impact only gets bigger when credit for the win is shared team-wide.

TREAT CONFLICT AS COLLABORATION

When disagreements happen on a team, people can get very territorial over ideas very quickly. That's a problem—because if defending your ideas involves destroying someone else's, then you're going to destroy respect alongside it. Instead, when conflicts arise and they stay task-focused (meaning conflict about team projects, shared tasks, or ideas for working better), teams meet that conflict as a form of collaboration. When someone speaks up to disagree with the consensus, they're doing so out of trust—not toxicity. (Except the four percent of the population who are sociopaths—but they probably need a different book). Team members who disagree (respectfully) are not saying the rest of the team is stupid. Rather, they're saying, "I care so much about the success of this project and this team that I can't help but point out a red flag I spotted or a different path we can take."

Resist the urge to push back on the idea. Try "Tell Me More" or RASA first. But if you must debate, be careful not to criticize the idea (and certainly don't criticize the person). Instead, start by questioning the assumptions behind the idea. Are they assuming a larger budget than we might have? Or are they assuming more people have time to work on their idea than the team really has? If you don't know what the assumptions are, try this simple question: what would have to be true for this idea to be our best option? In other words, tell me what you're assuming, and then let's go find out together. You may find out their assumptions were true, and that their idea is the best option. Or, you may find out they were way off. In either case, you'll have demonstrated respectful debate to the team and ensured that the person (and the whole team) is more likely to speak up in the future.

CALL FOR CANDOR

It's difficult to treat conflict as collaboration if no one on the team is speaking up. In that case, you may need to put out an open call for disagreement. Often, leaders *think* they solicit feedback. They'll put "Thoughts?" at the bottom of an email or end a discussion period during a meeting with "Comments? Concerns? Clarifications?" But many teams know these small openings are less like a real call for candor and more like the closing lines an officiant recites at a wedding. ("If anyone has any reason why these two should not be wed, speak now or forever hold your... comments, concerns, or clarifications.")

Instead, smart leaders put the burden of candor on themselves. They make it a request for help. So, if you feel like there's something going unsaid within the team—or if your team is heading to consensus a little faster than normal—put out a call to help the team by stating any disagreements. Say something like, "It sounds like we're in agreement, but that could be my own bias." Or, "Is there anything we're missing? Because I feel like I'm not seeing something." And then get comfortable waiting in awkward silence for someone to speak. Of course, when they do, model active listening and thank them for their honesty. Team members who really respect each other, and feel respected, know that candid criticism is their fastest way to growth—but some teams may need a little help getting to a place of regular candor.

AMPLIFY UNHEARD VOICES

On teams with strong personalities (or strong biases), certain team members will quickly begin to dominate the conversation. While we should model active listening, sometimes we'll hear softer voices trying to contribute and being ignored. In those moments, the least respectful action would be to allow the over-talkers to just keep dominating—which would signal to the rest of the team that you respect their voice more than others. Recall in Anita Williams Wooley's research in Part One on collective intelligence that the "smartest" teams made sure everyone spoke equally.

You may need to keep a list nearby and check off names as the more vocal team members share, or enforce rules around how long someone can speak. But it may also be as simple as noting who is most likely to cut someone else off and jumping in first to amplify and recognize what's being shared before they can shut it down. If the situation is really bad, consider a "pair and share" approach where team members are paired up for small discussions before coming back to the whole team to share. The twist is that people can't share their own ideas—each person must share their partner's ideas instead. (And if you're thinking of pairing the over-talker and the under-heard together so the dominator champions someone else's idea for a change... trust your instincts.)

When individuals feel respected—and respectful behavior becomes the norm on a team—trust will naturally increase, as well, ensuring that great ideas and great lessons get heard and considered. But for a team to move from a space where people feel understood and trusted to one where people feel they're doing their best work, they need to hear one more thing. They need to hear more about why their work is so important. As we'll learn in Part Three, the best way to get teams to hear "why" is to tell them more about "who."

PART THREE

PROSOCIAL PURPOSE

Common Understanding

Psychological Safety

Meaning + Impact

Prosocial Purpose

IN 2014, THE SENIOR LEADERSHIP OF KPMG FACED A dilemma.

Accounting was boring.

Apologies to any accountants reading this book, but that really was the dilemma. Bruce Pfau, KPMG's Vice Chair of Human Resources and Communication, had been working for nearly a decade to improve morale at one of the world's premier accounting and professional service firms. When Pfau took the position in 2004, "Morale was in the tank," he said.[57] Only about half of respondents to the annual survey said they had a favorable opinion of the firm.

Which is to say about half of respondents to the annual survey had an *unfavorable* opinion of the company they continued to work for.

They had made some progress over ten years. They increased benefits and compensation, and they created perks like more flexible work and more opportunities to advance, but the initial gains had leveled off. Hence, the dilemma.

One day, while pondering the problem and rereading the annual survey, Pfau zeroed in on a single question. It was a small item asking employees how much they felt their work had "special meaning." Its responses were pretty dull.

And it's easy to understand why. Accounting, and especially auditing, is a boring and often thankless job. Accountants spend most of their time staring at documents and spreadsheets while working out of the client's office, disconnected from most of their coworkers and connected to clients who don't really want to answer a lot of the auditor's questions.

In situations like that, it's hard to see the work you're doing as imbued with glorious purpose.

So, Pfau and his team decided to do something about that. They wanted their employees to know that the work they did truly was important. To communicate that, they did something surprisingly simple: They told stories.

At first, they told stories from the past. They launched a promotional campaign called "We Shape History." The title represented the desired answer to the question "What do you do at KPMG?" And to show how their firm shaped history, they collected stories from pivotal moments in world history in which KPMG had been involved.

They told the story of the United States passing the Lend-Lease Act, which authorized billions of dollars in aid to the allies during World War II. President Roosevelt had asked KPMG partners for help to oversee the logistics.

They told the story of how KPMG partners resolved $21 billion in conflicting financial claims, which laid the groundwork for the release of fifty-two American hostages from Iran in 1981.

They told the story of how KPMG certified the results of the 1994 South African presidential election and declared Nelson Mandela the winner.

They told stories that showed how KPMG's work did have special meaning, and they reinforced those stories through posters distributed throughout KPMG. Each poster had a different answer to the question "What do you do at KPMG?" like "We Reunite Families" for the Iran hostage crisis or "We Champion Democracy" for the election of Mandela.

The initial campaign was met with enthusiasm, which was good. Because Pfau's team was already planning on taking the movement even farther.

After the success of "We Shape History," KPMG launched the "10,000 Stories Challenge" in an effort to help employees personalize the way KPMG made an impact. This new campaign challenged employees to share how they made an impact in specific ways. In essence, "We Shape History" was about how KPMG had made an impact on history, and the "10,000 Stories Challenge" was about how KPMG was making lots of different impacts right now. They built an application online that could capture individual stories and help individual employees design a poster in the same style of the ones from the "We Shape History" campaign.

And the stories started pouring in.

KPMG employees shared stories and created posters with slogans like:

"I Combat Terrorism—because I help banks prevent money laundering that can go toward terrorist groups."

"I Help Farmers Grow—because I support the farm credit system that keeps family farms in business."

"I Restore Neighborhoods—because I audit community development programs that revitalize low-income communities."

Within three months, the application had captured ten thousand stories. Within six months, they had forty-two thousand stories.

Forty-two thousand ways individual KPMG employees felt their jobs had purpose.

Forty-two thousand ways those individuals made a meaningful contribution to the lives of others.

The campaign was an obvious success. And you could see it in the numbers. The next engagement survey saw massive increases in those who believed their work made an impact and that KPMG was a great place to work. The company jumped seventeen spots on *Fortune* magazine's list of the 100 Best Companies To Work For, becoming the highest of any of the "big four" accounting firms.

In explaining the success, Pfau said, "We helped them to elevate and reframe the work that they do. They could see themselves as conducting audits, or they could see themselves as protecting the life savings of the 53 million American families who had their life savings tied up in the equities market."[58] And while that was true about the company-wide initiatives, Pfau was most intrigued by what the data said at a team level.

During the first engagement survey after the campaign, Pfau's team added a simple question asking if the individual's manager regularly discussed *purpose* on their team. That information allowed Pfau to compare two different groups: one where teams really internalized and discussed purpose, and one where teams merely relied on the company-supplied media. The differences were stark. On the "discuss purpose" teams, twice the number of employees agreed that "I feel the work I do makes a difference." And around 30 percent more agreed with "KPMG is a great place to work." Pfau explained, "Employees whose leaders communicated about purpose were far more motivated to strive for continuous improvement and high performance than colleagues whose leaders failed to discuss this important topic."[59]

And that discussion not only engaged employees, but it also kept them on the team. Personnel on the "discuss purpose" teams were three times less likely to think about looking for another job—and also a third less likely to actually leave for a new company.

Taken together, two key elements stand out and help explain the success of KPMG's purpose initiative.

The first is that it wasn't just a corporate propaganda campaign. Senior leaders didn't just rewrite the existing mission statement, and they weren't content to just produce a series of motivational videos. They took the campaign to the individual and team levels, asking people to consider how their specific role made an impact (even though some team leaders foolishly ignored the help the campaign was designed to offer them).

The second, and likely more important element, is that KPMG's initiative was "prosocial." Instead of merely talking about their broad mission or their effect on an industry—and instead of merely talking about events from past history—KPMG's "10,000 Stories Challenge" was designed to focus on the individuals who were served by the work of specific teams.

When team members know they're making a meaningful contribution to work that serves others, increases in morale and performance are almost inevitable. Communicating purpose to teams is what we'll cover in the next two chapters. First, let's explore prosocial purpose a bit more.

What Purpose Really Means

To understand why KPMG's purpose initiative was so powerful, we need to talk about purpose. Most of us think about purpose as a bold and lofty ambition. Most leaders equate purpose with a mission statement. As a result, most teams fail to be truly inspired.

Think about the last time you felt engaged and motivated at work, or the last time you worked on a team that was inspiring and energizing to be a part of. You're probably not thinking about the last time your boss recited the company mission statement verbatim. Instead, you're probably thinking about the last time you got a "thank you" from a client or a coworker—the last time you or your team got to hear how the work you were doing mattered to someone else.

And if so, you're not alone. Over the past decade, there has been a rethinking among researchers on what it means to share a purpose that actually inspires. We all want to do work that matters. But since humans are social creatures, we tend to judge whether our work matters based on whether or not work matters based on who we can see being impacted by our work. We don't just want to know why what we're doing is important; we want to know, "Who is served by the work that we do?" That's because purpose is personal.

Purpose is personal.

Consider a study that Adam Grant and a team of researchers conducted in 2014. Grant and his colleagues were working with their university's donation call center. These call centers are manned by student workers who are given a list of alumni, along with a phone, and tasked to call each person and read from a script that always ends in a request for a donation. The persistent ones keep following the flowchart of the script and pair every "no" with a new request for a slightly smaller donation. (One time, I got one of these calls from my alma mater and was pestered all the way down to a request for $20.05 to commemorate my graduation year.)

The job is boring. It's draining to be hung up on, yelled at, or worse. It's relatively thankless. In fact, when Grant and his colleagues showed up, the first thing they noticed when touring the call center was a sign in one student's cubicle. It read, "Doing a good job here is like wetting your pants in a dark suit. You get a warm feeling but no one else notices."

The researchers wanted them to feel noticed—but obviously not for wetting themselves. They wondered if getting the call center employees to notice the difference they were making would have a motivating effect on them. So, they took the break time student workers received and used it to run an experiment. During a five-minute break, some of the workers were visited by a fellow student who had received scholarship funds raised by the call center, and they heard how receiving the funds positively impacted him.

They got to meet their answer to the question, "Who is served by the work that we do?"

And when the researchers followed up a month later, they noticed that just that small meeting with a scholarship recipient had made a big impact on the callers. The workers who got to meet the people directly served by their work worked twice as hard. They made double the number of calls per hour and spent twice the number of minutes on the phone. Their weekly revenue went from an average of around $400 to more than $2,000 in donations.

It's impossible to overstate how big this effect is.

The workers didn't get any additional perks or benefits. They didn't get any training. And they certainly didn't get asked to memorize and internalize the university's mission statement. Instead, they got a five-minute chat with someone whose life was made better by the work they were doing.

The researchers argued that these workers were inspired by a sense of prosocial motivation—the desire to protect and promote the well-being of others.[60] That term points to what's wrong with many organizations' attempts at talking purpose—there's nothing prosocial about it. When you're talking about growth, shareholder value, disruption, or even sustainability, it becomes awfully hard to also tell specific stories about specific people who are served by the work your teams do.

Prosocial purpose is what we want from our work and what bonds teams together. Prosocial purpose is felt when team members know they're making a meaningful contribution toward work that impacts others. It's about demonstrating the meaning behind the work teams are asked to do and then showing its tangible impact.

Meaning and Impact.

These terms may sound similar, but there are subtle differences that make each important. Meaning is knowing that your contribution counts. Impact is knowing who is counting on you. Taken together, they form the core of prosocial purpose. Prosocial purpose is more than just working for an organization that serves others. It's when everyone on the team knows how their work serves specific people.

In this final section, I'll explain what it means to show teams they're making a meaningful contribution, and we'll discuss how feeling the impact of their work can bond and motivate a team. We'll learn about meaning and impact by discovering the stories of some unlikely healthcare heroes—and an extremely minor league baseball team that makes a major impact.

CHAPTER FIVE

MEANING

WHEN YOU THINK OF HEALTHCARE HEROES, YOU THINK of doctors and nurses. We all do. We think of them running through the hallways of hospitals, treating the victims of tragedies, or protecting the vulnerable from deadly viruses. At the height of the recent pandemic, in cities around the world, people would gather on their balconies to cheer, clap, sing, and play music in gratitude for the brave doctors and nurses on the front lines. We think they are our primary defenders against the endless array of enemies who seek to harm our health.

But we rarely think of all the other people who keep us healthy.

Like medical assistants. Or receptionists.

We think of the doctor who listened to our heartbeat, or the nurse who drew our blood for testing. But we don't think of the woman on the phone who scheduled our appointment. Or the medical assistant who escorted us from the waiting room, took our vitals, and made us sit in our underwear on the cold exam table. But these jobs are equally important to keeping a healthcare system running.

But no one stood on their balcony and cheered for them. No one celebrates them as heroes.

Except at Kaiser Permanente.

For the past decade, California's largest healthcare system has been recognizing these often-unrecognized heroes and helping to put more meaning behind what's frequently seen as a thankless job.

It began without much fanfare, and even with a little re-sistance. Kaiser Permanente was in the midst of launching a unified and state-of-the-art electronic health record (EHR) system. Every single hospital, clinic, and doctor would be connected on one unified computer system that could track a patient through every corner of the healthcare system. EHR systems are more common today, but a decade ago, they weren't exactly warmly received. In Kaiser Permanente's case, the project took two years to fully install at an estimated cost of $4 billion. It required many of the 200,000-plus employees to be trained on how to use it—something few of those employees were likely motivated to do at the time.

But Kaiser Permanente's leaders saw an opportunity in their EHR system to motivate those same employees. For a long-time, Kaiser Permanente's leaders and many of their physicians had advocated for a focus on preventive care as the most powerful tool for improving health outcomes. Many of Kaiser Permanente's administrators and senior leaders knew this new system would be a powerful tool to improve preventative care.[61]

Alongside the launch of the system, Kaiser Permanente launched a program to celebrate collaboration across job functions and encourage everyone in every role to think of themselves as medical providers. The new system would allow anyone with access to pull a patient's record and see their entire history in one place, including whether that patient was up to date on preventative screenings.

When the system flags a patient as overdue for a screening or test, the staff member looking at the record is able to ask them about it and schedule them for the screening at that very moment. So, when a longtime patient calls her primary care physician about a cold, the receptionist can see if she's overdue for a mammogram and get that scheduled, as well. Or, when a new patient stands on the scale, silently debating just how much his shoes weigh, the medical assistant can see that there's no record of a colonoscopy on file and ask to schedule him for that during his check-up.

The program, called "I Saved A Life," empowers what are often considered support staff to think and act like health care providers. When one of those staff catches an overdue test, and the test finds a life-threatening illness, Kaiser Permanente considers that a life saved. And the local clinic or hospital celebrates that staff member's contribution with parties, certificates, and sometimes even pins or medallions engraved with "I Saved A Life" on them.

Since the program began, over 1,500 lives have been saved—which, by itself, is incredible. But what's equally incredible is how many staff have been reengaged and motivated in their jobs. It even motivated them to better utilize the EHR system. "Rather than staff seeing the EHR as a burden, they wanted to use the new system to save a life," recalled Robert Pearl, the former CEO of Kaiser Permanente.[62] Employees at Kaiser Permanente have reframed the "health care team" to include these previously excluded roles.

Many organizations suffer from the same delusion that the "I Saved A Life" campaign has cured. It's easy to believe that some jobs or some teams in an organization are more important than others. We use terms like "support staff" or "back office" to describe the people who help others do the "real work." But the truth is that every role in an organization is important.

It's all real work.

If it wasn't, then that work would have been outsourced to a different company long ago.

However, when leaders allow a caste system to develop between various functions and positions, the result can be a huge percentage of individuals and teams who are made to feel that their role has less meaning—or that their work is less meaningful to the mission of the organization.

The genius of "I Saved A Life" is the way a simple reframe around common tasks like scheduling and screening helped to restructure the entire organization into one where every position was seen as meaningful. Between the success of KPMG's "10,000 Stories Challenge" and Kaiser Permanente's "I Saved A Life" campaign, it's apparent that every team and every job can be seen as meaningful.

It just takes a team leader willing to help each person on the team see.

Why Meaning Matters

Most of us think of Meaning with a capital M. It's why we think of doctors and nurses as doing Meaningful work. They're saving lives. But the research on human motivation and team collaboration suggests something different. It's okay to offer lowercase m meaning as well. In fact, it's more than okay. Small m meaning dramatically increases big M Motivation for individuals and teams.

That research helps explain the brilliance behind Kaiser Permanente's "I Saved A Life Campaign." Receptionists and medical assistants may not have big M meaning. They're not reading the mammograms or giving the colonoscopies (that's probably a net positive, actually). But before "I Saved A Life," they didn't see how their day-to-day tasks related to those larger meaningful jobs and didn't see their contribution.

And contribution is the core of meaningful work.

Contribution is the core of meaningful work.

We often judge whether the jobs we have are meaningful based on how clearly we can see the contribution our work makes to the larger work of the organization. The most boring jobs are typically the ones that ask participants to do a seemingly random collection of tasks with little explanation about how the tasks create value for colleagues or customers. Not surprisingly, those jobs are often performed by the most bored and disengaged teams.

We want to know our work has a rationale behind it—a purpose, no matter how small.

And the lack of any rationale, or the lack of any lasting contribution, creates a lack of motivation.

Behavioral economist Dan Ariely conceived and led a rather creative study. Both the study design and the task it asked participants to do—build with Legos—tapped into creativity. Specifically, Ariely and his research team recruited participants to get paid to build Lego Bionicle figures. If you're too old (or too young) to know, Bionicles were these whimsical and weird action figures made from Lego parts that were sold only in the early 2000s. Ariely's team chose them because they were unique, but also because they reminded virtually everyone of building with Legos as a child—a nearly universal shared experience.

Participants were divided into two groups and offered two dollars to build a Bionicle action figure from the Lego parts. They were also told that any figures they built would have to be disassembled in order to be used by the next participant. In one group, after the first figure was completed, the researcher placed it out of sight underneath the table and then offered the participant more money to build a second one. But they offered eleven cents less money ($1.89 instead of $2.00). This continued for as long as it took for the participant to get bored or decide that building another one wasn't worth the additional money. Or both, really, because participants in this group built an average of eleven figures before quitting (for an average pay of fourteen dollars).

Participants in the second group were treated almost the same way, with the same Bionicle figures, same pay structure, and same cycle that continued until they quit. Except this time, rather than placing the completed figures under the table, the researcher would begin disassembling the first figure in front of the participants as they were building their second one.

Ariely and his team called the first group the "meaningful" condition, because those participants were allowed to see that they'd completed their work satisfactorily. Even though they were told the figures would need to be disassembled, they weren't reminded of it. By contrast, they called the second group the "Sisyphic" condition—after the Greek story of Sisyphus who was condemned to roll a boulder up a hill only to have it roll down over and over for all eternity. These participants watched their "hard" work get taken apart right before their eyes as they were working on the next task. As a result, they worked less hard. The Sisyphic participants only built seven figures on average—four fewer than those whose work had a bit more meaning.[63]

At first read, it's easy to overlook the study. After all, it's a bunch of college students building Legos for beer money. But the unavoidable lesson from the study is just how demotivating it can be to be asked to do work that doesn't matter. Even in two different conditions where the work didn't "really" matter, being reminded of that fact was even more demotivating. Another way to think of it is that *adding* meaning to a Sisyphic task can add motivation. And adding reminders about that meaning can add even more motivation.

Meaningful reminders can even turn individuals into better teammates. One of the most important forms of meaning is task significance. Task significance is knowing how your work enables others to work—it's knowing your work contributes to the ability of others to work. It's the receptionists and medical assistants being shown how their routine checks of patient history make a contribution to the work of saving lives from preventable diseases.

When people are shown their task significance, the quality of their work increases, and so does their willingness to help others do quality work—their willingness to be a good teammate. In one study, professors Adam Grant (yep, same Adam Grant from last chapter) and Francesca Gino found that individuals who helped out a colleague and then received "written expressions of gratitude" (the academic researcher term for thank you notes) were more likely to help the original colleague again, and also more likely to help other colleagues.[64] A thank you note represents a few things. It's a written expression of gratitude, sure. But it's also a reminder of how your actions helped someone else's work.

In a similar study conducted by Grant alone, lifeguards who were asked to read messages about other lifeguards who rescued drowning swimmers were more likely to stay focused on their jobs, and also more likely to help others they interacted with as part of their jobs.[65] This suggests that it doesn't even need to be a reminder of how *your* past work has had task significance and meaning—even hearing about how someone *else* doing the same job found meaning has a noticeable effect on helping behaviors—and, therefore, teamwork.

Taken together, all this research suggests that knowing our actions help produce a complete result, and knowing how that result fits into the larger team's work makes us more motivated but also more likely to be good teammates. So, it's no surprise that high-performing teams are more likely than mediocre teams to express gratitude to each other and help each other.

Making Meaning

At the same time, many teams struggle to see how their work fits into the larger picture. The larger the organization, the harder it is to see that fit, so it's easy to understand why so many individuals and teams are disengaged from their jobs. Fortunately, the research and examples like Kaiser-Permanente's "I Saved A Life" and KPMG's "10,000 Stories Challenge" prove that small interventions can have a massive impact on meaning. To get started, here a few simple ways to keep the meaning of your team's work at the top of your mind:

MAKE METRICS MEANINGFUL

Organizations are great at metrics. Or, perhaps better said, organizations that manage to stay operational are great at metrics. Measurements are what allow management to happen, and what allow teams to assess their performance. But metrics aren't meaning. Many leaders share performance metrics robustly. Senior leaders get excited to share just how much the company has grown, how much more revenue it has taken in, or the clients or stakeholders it has served. Team leaders enthusiastically share how the team is performing using the key performance indicators the organization uses to judge the team. (Or, sometimes not so enthusiastically, they share how much the team is lagging behind).

But often the blur of trackable metrics makes it difficult to remember why those metrics matter. High-performing teams know their numbers, but they also know why those numbers matter—and if they don't, they have a leader who can readily remind them. Consumer goods company Péla is on a mission to create a waste-free future by creating everyday plastic goods out of a compostable plastic alternative. While they track sales and revenue, the most important number stays front and center: how much plastic they've kept out of the ocean. Even their customers are reminded of their meaningful metric. If you buy an item from their online store, you're immediately shown a number representing the equivalent number of plastic bags you've kept from being trashed. (If you're wondering, Péla has currently saved the equivalent of roughly fifty million plastic bags from entering oceans—a number that gets shared right alongside every sales report and profit and loss statement.)

GIVE THE "IT'S A WONDERFUL LIFE" TEST

One powerful thought experiment for finding or reinforcing the meaningful contribution the team makes is the *It's A Wonderful Life* test, named after the black and white film synonymous with Christmas (despite it originally being a box office flop). In the film, an angel stops a depressed George Bailey from jumping off a bridge and shows him what his community would look like if he weren't a part of it. Don't worry—we'll stay a lot less morbid. But the thought experiment remains the same.

Have the team try to imagine how the world would be different if your organization didn't exist. How would the industry, society, or community be different? Then, take it a level deeper and ask the same question of your team. How would the organization be different if your team didn't exist? Talk about the void that would be left if your team suddenly disappeared. Just like in the film, realizing the impact of your absence is a great way to meaningfully connect with the impact of your continued presence.

CREATE TEAM SYMBOLS

Symbols are powerful reminders of the meaning and purpose behind a team, tribe, culture, or company. They're the visual embodiment of your team's mission and priorities. The process of selecting a symbol can be an equally potent activity for connecting your team and getting them to find the meaning in their contribution. The symbol doesn't have to be fancy, and it doesn't have to be an elaborate branding exercise. It can be a picture, a statue, a piece of clothing, or even an action figure or doll. It can be one big symbol, or something small and cheap that can be handed out often.

Years ago, I spoke at the US Naval Academy, and afterward, the officer who invited me presented me with a small coin as a token to remember my time there. When I asked him more about it, I learned about the tradition of challenge coins used in various branches of the US military to commemorate important missions or reinforce the shared identity of a certain division. There's no bureaucratic rule about how these coins are designed or presented; instead, each team and each leader decides what goes into it—and in that way, it becomes a lasting symbol of that team and its contribution. In the corporate world, I've worked with companies that have used everything from expensive trophies to cheap stress balls as symbols to remind their teams of the larger mission and how their work brings the whole organization closer to it.

CREATE A RALLYING CRY

A symbol may be worth a thousand words, but only when people know what those words represent. So often the best teams pair their symbol with a rallying cry (or forgo the symbol altogether in favor of their chosen call to arms). It may be tempting to dismiss catchphrases or calls to action as corny or inauthentic, but cringe may be the source of their power. Rallying cries create clarity and summarize in a short sentence the work a team does and the difference that work makes. They're clear, crisp reminders about why the team's work matters.

The San Antonio Spurs adopted the rallying cry "pound the rock," tapping into the imagery of a stone-cutter hitting away at a rock hundreds of times before it finally cracks. The Spurs use it as a powerful reminder that the day-to-day strain of training and drilling hundreds of times is what brings victory. At their practice facility, they tie the rallying cry to a powerful symbol: a boulder and sledgehammer displayed behind glass. The New Zealand All-Black's encourage each other to "leave the jersey in a better place" to remind players that it's not about them, but about the team and the community it serves. And KPMG's "10,000 Stories Challenge" was an invitation to each individual and team to draft a rallying cry for how the work they were doing made a difference. Ask your team "who is served by the work that we do?" and then craft their answers into your own rallying cry to keep serving.

SHARE A WIN EVERY DAY

Most teams celebrate wins, but they're often limited to the successful end of a project or hitting an important milestone. High-performing teams share wins much more frequently. In fact, many teams try to close out every day with a chance for each member to share a small win. It may sound like that's taking too much time for something of too little importance, but that's exactly the point. When teams are bogged down with the small tasks that make up the day-to-day experience, it's hard to keep meaning and contribution at the front of their minds. But closing every day with a win gives them something to celebrate—and a small reminder that they're making progress on work that matters.

And it's not just reflecting on the small win personally that makes a difference. When team members share those small wins publicly, the round of applause and encouragement that comes from their peers reinforces how valuable their contribution to the team is. Depending on the team, the method and timing of shares might vary—but don't let that make people think their win is too small to share. The more frequently teams share, the smaller the wins that get shared. Counterintuitively, the smaller the wins shared, the bigger the impact they have on the team.

Meaning is about seeing—or helping the team see—how their work matters and how it makes a contribution to the larger organization, its customers, or its stakeholders. And it's an important first step in keeping a team bonded and motivated. But it's easy to speak in grandiose platitudes about mission alone. Teams do want to make a meaningful contribution to the mission, but they also want to know how that mission serves specific people. Showing your team their specific impact is the subject of our final chapter.

CHAPTER SIX

IMPACT

THE SAVANNAH BANANAS ARE THE GREATEST BASEBALL team you've never heard of. Or, maybe you *have* heard of them. Maybe you're one of the millions of fans following their crazy dances and silly antics on their social media accounts. Maybe you're one of the thousands attending their sold-out shows that feel more like a circus than a baseball game. If you have heard of them, you've probably found yourself asking the same question everyone asks right before they relax and decide to become a fan:

"How is any of this possible?"

The short answer is: it shouldn't be. The Savannah Bananas shouldn't exist. Even if they did, they shouldn't have found anywhere near the success they've had.

First, let's consider exactly where they fit into the baseball hierarchy.

At the top of the baseball pyramid is Major League Baseball—the only league most people even know of or pay attention to.

Then there's the minor A-leagues, farm teams for the major league but based in mid-tier markets. Triple A at the top, then Double A, then Single A, and then Rookie league.

Then, all the way below Rookie league, there are the college baseball leagues like the Cape Cod League, the Alaska Baseball League, and the Coastal Plain League. These teams fill their rosters with college players on summer break.

And it's in the Coastal Plain League, a fifteen-league team based in the American southeast, that the Savannah Bananas founded BananaLand—their nickname for the only stadium in college baseball that actually fills up with fans every game.

The Savannah Bananas are the brainchild of Jesse Cole, a college player destined to pitch in the pros before a shoulder injury ended his playing career early. Undaunted, Cole found a different career working first as the General Manager of the Gastonia Grizzlies (the idea for the Bananas wouldn't fully ripen for a few more years). The fact that a twenty-three-year-old former player with no management experience was given a General Manager position should have been a warning about the status of the team.

But Cole started experimenting with creating more memorable experiences. He started choreographing dances for the players during lulls in the game (something that happens far too often in baseball). He organized a Grandma Beauty pageant in partnership with a local nursing home. He put a dunk tank outside the ballpark and invited spectators to try to dunk the crazy General Manager sporting a yellow tuxedo: him.

Little by little, Cole's experiments turned a Grizzlies game into something people actually wanted to watch. The team that had averaged 200 fans a game before he took over was selling out their 2,000-person stadium. Cole had made an impact in Gastonia but wondered if he could make an even bigger one somewhere else.

During a trip to Savannah with his fiancée (whom he'd met working with the Grizzlies and convinced to marry him despite his constantly wearing a ninety-nine-dollar yellow tuxedo), Cole attended a game at Grayson Stadium, a ninety-year-old stadium that was home to a Single A minor league team called the Savannah Sand Gnats. "It was a Saturday night," Cole recalled. "And we walked into the stadium to the grandstand, and there was no one there."[66] It was a beautiful night for baseball, but the city of Savannah didn't seem to agree.

That gave Cole an idea. "I called the commissioner of our league and said, 'If the Savannah market ever becomes available, we've got ideas.'"

The commissioner likely thought he was crazy. But within months of that call, the Sand Gnats announced they were relocating—which made them the ninth team in a row to have tried, and failed, to make it work at Grayson Stadium. Cole thought his results would be different. And why not? He'd learned to do just about everything in baseball differently.

When Cole announced their intention to launch a brand new team in 2016, the city likely thought they were crazy. And when they announced their name a few months before opening day, the city *knew* they were crazy. The Bananas became the number one trending topic on Twitter, mostly because of Savannah residents who hated the name. They posted messages like "This is an embarrassment to the city of Savannah," and "You'll be giving away tickets to keep up the morale of the team." But there were a few glimmers of hope. A swell of support was building just beyond the online trolls. The Bananas name announcement ran on several national media outlets, including ESPN's *SportsCenter*. The team sold out of T-shirts within twenty-four hours of the announcement.

At the center of Cole's strategy was putting fans first (in fact, Fans First Entertainment is the actual name of the company operating the team). "We exist to make baseball fun," he said. "It's a dying sport." Major league games are long, drawn-out affairs with too many breaks, too many advertisements, and too many upsells. "You have to pay surcharges to buy tickets. You have to pay for parking. You even have to overpay for underwhelming food." The Bananas would do things differently. One ticket price included everything—seats, parking, and even food. To minimize the boredom caused by breaks, the Bananas would put on a show.

It started small, with players doing dance routines or playing weird games, or selecting a child from the crowd to be the "banana baby" and lifting him up amidst a circle of kneeling players while the opening song from Disney's *The Lion King* played. (That last one isn't really that small, actually.) But it grew from there. To continue to entertain the fans, the Bananas added a breakdancing first base coach, and then had players introducing themselves during walkout routines that rivaled the hype of professional wrestling. Eventually, they'd go even further for the fans. When spectators entered Grayson Stadium—accompanied by the parade of a marching band—both teams playing that night would line up to give them high fives as they entered. After the game, Bananas players would throw an after-party and sign autographs until every fan left the stadium. At almost every interaction, Bananas players and staff said the same thing: "Thank you for coming out tonight."

It worked.

In their first season, the Bananas sold out seventeen of their twenty-five games. In their second season, they sold out every game. That began an unbroken streak of sellouts that now numbers more than 150 games in a row. Every year, they set a new record in attendance for the Coastal Plain League and draw crowds that would make even Triple A minor league teams jealous.

And success isn't judged by ticket sales alone. The Bananas win games—a lot of them. They won the Coastal Plain League championship their first season. They won it again in 2021. And they've finished near the top of the rankings in every other season they've played. The Bananas have won more games since they started than any other team in the league.

"Our players literally play better when they're with our team," Cole boasted. In 2019, Georgia Southern University professor Curtis Sproul analyzed the records of each Bananas player and compared them to their records when they played for their college teams during the school year. When he finished the calculations, Sproul found that when they played for the Bananas, players had a significantly higher on-base plus slugging percentage (a nerdy metric of offensive ability). In fact, the Bananas team was the only collegiate summer league team where players performed better than when playing for the college team.

Cole has a simple explanation: the culture. Players are having fun and playing for the fans. Unlike a lot of college and minor league teams, Bananas aren't focused on their development and their potential to be pros. Unlike even the major league teams, Bananas aren't focused on their performance and their paychecks. Instead, they're focused on the fans. Every single night, players have multiple interactions with fans—multiple chances to hear about how they're impacting the lives of spectators through their wild and crazy version of baseball.

In 2021, as the team prepared for the league championship once again, Cole knew it was the impact that would motivate them more than anything else. In the days leading up to the championship series, Cole reached out to fans. On the first night of the series, instead of a rousing motivational speech, Cole and the coaches just played a video—a compilation of fans sharing how much joy the team had brought them.

The Bananas took the field knowing they weren't playing for themselves. They were playing for the fans. That fans-first mentality brought them the championship. In 2022, the Savannah Bananas again won the league championship, becoming champions in three out of the six years they'd been in the league.

And shortly after winning, Cole announced it would be their final championship. The Bananas were leaving the Coastal Plain League and becoming a professional baseball team that could pay their players and take their show on the road more often to entertain even more fans. Now, the Bananas will play twice the number of games and impact that many more people.

Because it was never about winning; it was always about the fans. Never about what or why, but always about who.

How We Actually Measure Impact

The Savannah Bananas' motivational strategy is so brilliant, and it serves as such a potent example of prosocial purpose, because it emphasized the impact the team made on specific people. Their pre-championship video, like everything they do, served to remind the players that they exist to create and serve fans.

But their strategy also points to what's missing from the work life of so many employees. Many companies have mission statements they can point to, but few can point to specific people who are served by the company—and fewer still can do it on an individual or team level instead of across the organization. When people are mentioned, it's often with the incredibly vague labels of "shareholder" or the increasingly popular "stakeholder." We know purpose is important. We know people want to make an impact. But we fail to give people purposeful work because we're talking in terms too grandiose or generic for your average team. Perhaps that's why a 2019 study from McKinsey found that 82 percent of employees agreed company purpose was important, but only 42 percent felt their organization's purpose statement actually made an impact.[67]

We don't measure impact with the flowery prose of a mission statement. We measure it by the people whose lives or work we see changed.

Purpose isn't a *why*.
Purpose is a *who*.

But many organizations' mission statements are missing that *who*. In 2020, professors Christopher Michaelson, Douglas Lepisto, and Michael Pratt asked nearly two thousand CEOs to describe the purpose of their organization. More than half failed to mention any beneficiary at all. (And yes, many of those who did merely mentioned shareholders.)[68] Think about the missed opportunity for impact, or the impact that awareness of their influence would have on employees. One of the core questions of Gallup's famous Q12 employee engagement survey asks whether the mission or purpose of the company makes employees' jobs feel important. And yet, decades after Gallup launched the survey, engagement numbers are still hovering in the same range they've always been at, of 20 to 30 percent of employees indicating they are engaged.[69] It's no wonder why.

However, the real power of making a prosocial impact—of knowing who is served by the work you're doing—is best seen on an individual and team level. We've already seen how a brief visit from a student who received scholarship funds raised motivation and performance in the call center workers tasked with raising funds. But similar boosts have been found in other fields. In one study, placing a photograph of the patient next to their set of X-rays boosted the accuracy of readings by radiologists— even though they'd never met the patient.[70] (Despite what medical television dramas show, the doctors performing tests and X-rays and the ones reading them are almost never the same.) In another study of physicians, placing signs near hand-washing stations with the message "Hand Hygiene Prevents Patients from Catching Diseases" led to 45 percent more soap and hand sanitizer usage (and, presumably, more hand washing) than similar signs focused on preventing oneself from catching disease.[71]

The missing link between an individual's prosocial motivation and an organization with a truly inspiring purpose appears to be at the team level. It's not about what flows from the top; it's about what's being discussed in the middle. Professors Claudine Gartenberg, Andrea Prat, and George Serafeim surveyed around 500,000 employees at various United States-based companies to learn their perceptions of corporate purpose and see how it correlated to publicly available financial performance. Unsurprisingly, firms whose employees didn't feel much purpose in their work had worse financial results than firms with purpose-driven employees. But surprisingly, what drove the majority of employees' perceptions of purpose wasn't senior leadership, but rather middle managers discussing and emphasizing purpose. The survey was a 500,000-person, multiple-company replication of what KPMG found internally after implementing their "10,000 Stories Challenge."[72]

All of this research, plus examples like KPMG, Kaiser Permanente, and even the Savannah Bananas suggests it isn't just that individuals want to know they're working for a firm that makes an impact on others—they want to know how *they* and *their team* are making an impact. The best-performing teams don't ask individuals to join the organizational purpose; they exist in organizations willing to help them find their individual purpose—and to see their impact.

Seeing the Impact

At first glance, letting teams see the impact of their work should be easy. Serving customers or stakeholders is the goal of every organization. But as companies grow larger and more complex, more people find themselves in roles where they don't get to see those customers or stakeholders often—and so impactful reminders become even more important. Fortunately, there are several tactics you can adopt from teams and companies to keep impact front and center:

COLLECT IMPACT STORIES

Stories are the most powerful influence on our emotions and sense of purpose. Stories light up our brains and help us make sense of the world in a way that only a millennia-long tradition can. Research suggests that stories are twenty-two times more likely to stay in our memory than facts or raw data alone. (It's the reason every chapter of this book opens with a story.) So, if you want to help your team remember the impact they make, tell stories. Particularly, share Impact Stories, which help the team build a mental bridge between the work they do and the people who receive their contribution. Answer "Who is served by the work that we do?" with a story. Anytime you can, capture specific examples of your team's positive impact on colleagues, clients, or the community. You may want to go so far as to create a folder in your email inbox to save stories so they're already at the ready. When someone emails you a "thank you," save it in the folder. When you hear an anecdote about how your team's work helped, email yourself a quick note and save it in the folder. Then, any time you're kicking off a team meeting, you have a plethora of stories to draw from to create a powerful opener.

It may sound odd to lead with a story, because stories are the intangible aspect of management. And the intangible nature of stories is what creates the emotion and the impact. You can't measure a story. But, you also can't *feel* a metric.

PAUSE FOR PURPOSE

If you asked employees when they are furthest removed from feeling the impact they make on others, the answer would be nearly unanimous: meetings. When dealing with vague agenda items, overly-worded PowerPoints, and that guy who spouts opinionated statements that change pitch at the end to make them seem like questions, it's hard to keep "who is served" at the front of everyone's mind. So, before starting a meeting with your team, take a pause for a moment to highlight "who" and talk about the real purpose behind the work you're doing and how it serves others. It doesn't take much. Share just a few short sentences about how excited you are to be part of a team making an impact. Grab an email from the Impact Story folder and read it aloud. Or, make like Amazon and leave an empty chair at the meeting room table to remind everyone that, even though the customer or beneficiary isn't in the room, they still matter—and decisions made should serve them.

Purposeful pauses don't have to be solely for meetings. They also work well just before the team gets to work. At Beth Israel Deaconess Medical Center, the entire team of surgeons, nurses, and support staff pause before every surgery and take a moment to remember the patient they're about to operate on. They break up what would be a routine procedure with a powerful reminder of the humanity behind the work they're doing.

OUTSOURCE INSPIRATION

We've already covered how purpose discussions reso-
nate better when they happen at the team level, led by
team leaders instead of senior executives. But what's even
better than hearing from your teammates and team leader?
Research suggests people are most inspired when they hear
about the impact they're making from the very people
they're impacting, like the call center study that opened
this section. A growing number of organizations are out-
sourcing inspiration in innovative ways. Medical device
company Medtronic invites patients to the annual holi-
day party to share how the company's technology helps
them live better. Best Buy created a service they called
"Twelpforce" where employees in all roles respond to customer questions and inquiries on Twitter. In its first year,
more than 2,600 employees signed up, including employees
who didn't normally get to interact with customers. At
Whole Foods, even non-customer-facing roles like inven-
tory and restocking become customer-facing, since employees are encouraged to educate shoppers about organic
food standards, sustainable food production, and cooking
methods—often running classes in the store.

But the award for most committed and most complex
way to outsource inspiration goes to The Lego Group.
Since Lego's purpose is to entertain and promote creativity in children, they outsource inspiration to...children.

Each week, Lego brings in a new group of kids to "test" new products and share their experiences. Employees get valuable feedback on how to improve the product—but, more importantly, they get to see and share the experience of their product, directly enhancing the creativity of their most valuable end user.

BECOME THE BENEFICIARY

While some companies outsource inspiration by bringing customers in, others reinforce impact by sending employees out. They put employees in the role of customers in order to better understand the impact they're making and how to improve. Four Seasons Hotels include a "familiarization stay" in their employee orientations. Housekeepers and hotel clerks are given a night's stay in the hotel to better understand the service experience.

And the Savannah Bananas invite non-player members of the team to become "undercover fans." Team members get a general admission ticket and wear regular clothes. Similar to Lego bringing children in, Bananas team members get insights on how to improve the experience of attending a game, but they also get an up-close look at how the game impacts attendees—and get a chance to capture stories from fans they might never have heard otherwise. Whether it's playing the role of a customer or stakeholder, or hearing from them, a powerful way to understand "Who is served by the work that we do?" is to experience that impact as directly as possible.

In the end, people want to do work that matters. And teams work best when they know who is served by that work. Learning to talk about prosocial purpose with a team provides the sense of meaning and impact that mission statements and corporate propaganda pieces cannot. It may be the final of the three elements of a great team culture, but it's often the one most teams need to start with. Once teams share a specific answer to how their specific work makes a difference, they're motivated enough to build more common understanding and psychological safety. As we'll see in the conclusion, they'll also start finding all three elements yield more positives than they first expected.

CONCLUSION

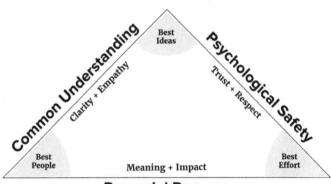

WHEN GARRY RIDGE WAS APPOINTED THE CEO OF THE WD-40 Company, the company was in a uniquely difficult position. It was a beloved brand, whose signature blue and yellow cans (with the little red top, and tiny red straw) were in 80 percent of American households—and virtually 100 percent of factories, construction sites, and repair shops. It had grown to take over its entire market. But as a result, the company wasn't growing anymore. As an analyst for *Barron's* put it, "the very nature of WD-40's past success doomed it to ultimate failure."[73]

But Ridge had a plan to escape that failure. He wouldn't focus on improving the company's product (although that would come in time). Instead, he'd focus on improving the company's people. He'd work on building a better culture on the team—or, as he called it, the tribe.

Ridge became known for spending the first two hours of his workday with employees—excuse me, with tribe members.[74] He took the time to greet them as they came into work, leveraged time spent doing nothing to build connections and bonds with them, and helped them do likewise with each other. Over time, this helped to build the empathy part of what we'd call common understanding.

But it wasn't just a touchy-feely connection. Ridge also focused on building a culture of clarity on individual teams and throughout the entire organization. As part of getting WD-40 Company tribe members to act as part of a true tribe, Ridge launched a concept he called the "Four Pillars of the Fearless Tribe." Those pillars were Care, Candor, Accountability, and Responsibility. We'll come back to care and candor; but for building common understanding, accountability and responsibility were huge.

Accountability, according to Ridge, didn't mean managers holding people accountable when they fell short of some standard of performance. "We see it as a two-way street in which leaders and their direct reports equally hold ownership of the way we perform our duties and what outcomes our efforts lead to," Ridge said. Accountability was about leaders making sure their direct reports knew exactly what was expected of them, and giving them everything they needed to deliver on those expectations. Or, as he sometimes phrased it: "Don't mark my paper; help me get an A."

Responsibility to the tribe meant accountability to the whole team. It meant taking personal ownership of expected outcomes, because the tribe was counting on each individual to deliver those outcomes. Responsibility happened when tribe members understood their role in the tribe's desired goals and felt responsible for supporting the tribe by performing in that role. "Everyone has their part and role," Ridge said. "And everyone performs exactly as expected, because everyone shares the responsibility of a successful outcome."[75]

That might be one of the better ways to describe common understanding.

But turning around The WD-40 Company wouldn't just require accountability and responsibility. It also required creating more psychological safety (what Ridge called "care and candor"). Ridge knew that fear was one of the most disabling emotions humans felt, and that many people felt fear at the hands of their team and managers far too often. In addition, trying to emphasize accountability and responsibility would be impossible if fear of failure still lingered in the tribe. So, Ridge focused on reframing failure. "Around here, we don't have mistakes," he announced. "We have learning moments. A learning moment is a positive or negative outcome of any situation that is openly and freely shared to benefit all."[76] After all, The WD-40 Company was a company born out of learning moments. The name is short for "water displacement formula 40," because it took inventor Norm Larsen that many learning moments to get the formula right.

In the WD-40 Company tribe, anyone could openly share a failure without fear of punishment if they simply announced, "I had a learning moment, here's what happened, and here's what I learned from it." Learning moments weren't failures. They may have felt like frustrations—but because they carried with them hard-earned lessons, they were victories, as well. "My job is to create a company of learners," Ridge said. He knew that if the company was going to find new and innovative ways to market WD-40, then tribe members were going to experience failures. His job would be to help them admit and share those failures. Not everyone took to the idea of learning moments, so Ridge did something drastic. He encouraged any one of the hundreds of tribe members to email him their learning moments directly, and every month he would award a prize to the best one. The first month didn't see a lot of submissions, but the ones that did come in allowed Ridge to reinforce the importance of learning moments. "They saw that I was serious, that a learning moment was to be applauded and rewarded," Ridge said.[77] Over time, the number of submissions grew and grew.

As team members became more comfortable sharing failures, they became more comfortable sharing all sorts of things. Remember, learning moments weren't strictly a euphemism for failures. According to Ridge, the moments were to include any positive or negative outcomes, so long as they were shared for the benefit of all. Getting tribe members to see successes and failures as equal lessons, in terms of learning, meant more people playing the role of teacher and student at the same time. It meant more people building trust and a desire to share their insights and respect for each other's hard-earned brilliance.

In other words, it meant a culture of psychological safety.

But what about prosocial purpose? After all, WD-40 is literally just oil in a can. How inspiring can selling more cans of oil be? To Ridge and the tribe, it's very inspiring.

"Our products make heroes of people!" Ridge boasted. "If you have a squeaky door, it's driving your family crazy, you solve that, and everybody loves you."[78] Ridge told anyone who would listen just how transformative a simple blue and yellow can of WD-40 can be. If you've ever dealt with a door hinge that won't stop squeaking, or a kitchen drawer that keeps getting stuck, you know how transformative a few squirts of WD-40 truly are. You may have even become a hero to your own tribe in the process. Ridge got his tribe to see that most households keep a can in their house for the same reason Superman keeps his super suit underneath his clothes: you never know when you'll need a hero.

Over time, reframing WD-40 into a blue and yellow can of purpose also helped the tribe rethink ways to help their hero customers solve more problems. They launched new ways to package their proven formula—new ways to apply the oil to new problems. They pushed into new markets, expanding from an American company to a truly global one. In fact, under Ridge's tenure, the company went from drawing only 20 percent of its revenue from outside the United States to drawing 65 percent of revenue internationally. In the same period of time, The WD-40 Company delivered a compound annual growth rate in excess of 15 percent to its shareholders.

But it's what Ridge delivered to the tribe that matters to him more. When Ridge took over, The WD-40

Company was a mediocre company with a product many experts felt was at the end of its life cycle. Ridge transformed the culture to one of common understanding, psychological safety, and prosocial purpose—and created a company employees and customers loved even more than they thought was possible.

He set out to build a tribe, and in the process made the team at The WD-40 Company the best it ever was.

Building the Best Team Ever

What Garry Ridge did at The WD-40 Company is actually how culture transformation works on most teams.

Throughout this book, we've discussed the three elements of a high-performing team culture as just that: elements. We've reviewed common understanding, psychological safety, and prosocial purpose as separate characteristics. We broke each one down into its component parts, which technically demonstrates they're not elements in the chemical sense.

But, in a way, they work precisely like chemical elements—because they aren't standalone characteristics.

They're interdependent. If one changes, that affects the others.

And they interact when they combine and take on new features.

When teams start working on more than one element, they notice changes they weren't expecting.

When you build common understanding and prosocial purpose, you start to attract more people to your cause, and more talented people join the team. That's because people are doing work that matters alongside people they deeply care for and enjoy collaborating with—just as the UCLA women's gymnastics team recruited more and more top performers once Valerie Kondos Field led from a place of empathy and a different perspective on winning than the rest of the sport.

When you build common understanding and psychological safety, those same people start to generate their best ideas, and you can solve more complicated problems and create more value. When those two elements combine, people feel they can truly express themselves and understand the full expression of their teammates—just as Maggie Wilderotter's culture change at Frontier led to a better understanding between white-collar and blue-collar workers; and better solutions for how to serve customers, resolve union disputes, and grow the company.

And when you build psychological safety and prosocial purpose, you start getting the best effort from your team because they know they can experiment and take risks, and they know how their work makes an impact—just like the Savannah Bananas did when they gave their players and staff room to experiment in the service of serving fans.

When you build common understanding, psychological safety, and prosocial purpose, these elements combine and interact, and you start to make something powerful.

You start to create the best team ever.

Or, at least the best team you've ever led.

JOIN THE TEAM

Spread the Word

Liked *Best Team Ever*? Help us spread the word by writing a review for others to see. It's a small action, but it makes a big difference to those considering the book.

We've outlined the simple steps to follow at davidburkus.com/review.

Grab the Special Resources

If you want to go further in building your best team ever, we've created a collection of extra resources, including playbooks, worksheets, video training, and more.

All of these resources are freely available at davidburkus.com/resources.

Bring in David

And if you want to go *even* further, invite David to your next company meeting, retreat, conference, or event and let him inspire your team to do their best work ever. David will delve deeper into his research in a way that is tailored to your organization and industry, helping everyone walk away with a practical plan of action for building their best team ever.

Learn more at davidburkus.com/speaking.

ACKNOWLEDGMENTS

IT FEELS LIKE BEST TEAM EVER WAS WRITTEN ALONGSIDE the best writing team I ever had.

The team at Scribe Media who helped launch both this book and Twinbolt: Tucker Max, JeVon McCormick, Chas Hoppe, Emily Gindlesparger, Hussein Al-Baiaty, Nicole Jobe, Tom Lane, and Lily Wood. And my most trusted advisors on the project: Clay Hebert, Laura Gassner Otting, and Bret Simmons.

The amazing team leaders who made themselves available for interviews: Jesse Cole, Matt Bertulli, Valerie Kondos Field, Garry Ridge, and Alan Mulally.

The researchers whose work on teams and team culture whose work has illuminated our understanding: Dan Ariely, Peter Bamberger, Henrik Bresman, Christopher Chabris, Carol Devine, Giada Di Stefano, Amy Edmondson, Amir Erez, Trevor Foulk, Francesca Gino, Adam Grant, Nada Hashmi, Emir Kamenica, Tami Kim, Kevin Kniffin,

Amir Kugelman, Simon Lam, Thomas Malone, Michael I. Norton, Alex Pentland, Gary Pisano, Ann Chunyan Peng, Christine Porath, Dražen Prelec, Jane Risen, Arieh Riskin, Kinneret Riskin, Julia Rozovsky, John Schaubroeck, Juliana Schroeder, Ovul Sezer, Irit Shoris, Jeffery Sobal, Bradley Staats, Brian Wansink, Anita Williams Wooley, and Paul Zak.

The thought leaders who built on that research and made it accessible: Shawn Achor, Chester Elton and Adrian Gostick, Ben Horowitz, Daniel Coyle, Shane Snow, Heidi Grant, Jon Gordon, and Patrick Lencioni.

The early reader team whose feedback helped make this the best book ever: Kevin Anselmo, Maggie Bolden, Donald Dunnington, Keita Demming, Steve Goble, Kathryn Haydon, Theresa "TJ" Jump, Susan Litwiller, John Spence, Tathra Street, Harrison Wendland, Nikita Zhuk, and Jennifer Zimmerman. And the team that created our early reader platform, Help This Book, including Rob Fitzpatrick, Devin Hunt, Marc Davies, and Adam Rosen.

And my wife, Janna, and our two boys, Lincoln and Harrison...my best team ever.

ENDNOTES

1   Jason Hahn, "U.S. Men's Curlers Make History with an Olympic Gold-Medal Win—After a Pep Talk From Mr. T," People, February 24, 2018, https://people.com/sports/winter-olympics-2018-team-usa-men-curlers-gold-medal-win/.

2   Adrian Gostick and Chester Elton, The Best Team Wins: The New Science of High Performance (New York: Simon & Schuster, 2018), 1.

3   Chris Hadfield, An Astronaut's Guide to Life on Earth: What Going to Space Taught Me about Ingenuity, Determination, and Being Prepared for Anything (New York: Little, Brown and Company, 2013), 82.

4   Hadfield, An Astronaut's Guide, 103.

5   Jim McNally, "Astronaut Marshburn's Mother Dies," Hickory Daily Record, last modified March 20, 2019, https://hickoryrecord.com/

astronaut-marshburns-mother-dies/article_260ae1d4-98c4-11e2-ab18-0019bb30f31a.html.

6 Anita Williams Woolley et al., "Evidence for a Collective Intelligence Factor in the Performance of Human Groups," Science 330, no. 6004 (2010): 686–688, https://doi.org/10.1126/science.1193147.

7 Anita Williams Woolley et al., "Bringing in the Experts: How Team Composition and Collaborative Planning Jointly Shape Analytic Effectiveness," Small Group Research 39, no. 3 (June 2008): 352–371, https://doi.org/10.1177/1046496408317792.

8 Diane Coutu, "Why Teams Don't Work," Harvard Business Review, May 2009, https://hbr.org/2009/05/why-teams-dont-work.

9 Donna Hood Crecca, "Interactive Education: Pal's Sudden Service Combines Technology with Hands-on Training to Drive Results," Chain Leader, February 1, 2005, http://maamodt.asp.radford.edu/Aamodt%20(5th)/Case%20Study%20Articles/Case%20study%20-%20Pal%27s%20Sudden%20Service.pdf.

10 William C. Taylor, Simply Brilliant: How Great Organizations Do Ordinary Things in Extraordinary Ways (New York: Portfolio/Penguin, 2016), 47.

11 Gary P. Pisano, Francesca Gina, and Bradley R. Staats, "Pal's Sudden Service—Scaling an Organizational Model to Drive

Growth," Harvard Business School, Case 916-052, May 18, 2016, 9, https://www.hbs.edu/faculty/Pages/item.aspx?num=51173.

12 Pisano, Gina, and Staats, "Pal's Sudden Service," 9.

13 Lynda Gratton and Tamara J. Erickson, "Eight Ways to Build Collaborative Teams," Harvard Business Review, November 2007, https://hbr.org/2007/11/eight-ways-to-build-collaborative-teams.

14 Daniel Coyle, The Culture Code: The Secrets of Highly Successful Groups (New York: Bantam Books, 2018), 229.

15 Heidi Grant, "Get Your Team to Do What It Says It's Going to Do," Harvard Business Review, May 2014, https://hbr.org/2014/05/get-your-team-to-do-what-it-says-its-going-to-do.

16 Valorie Kondos Field, "Why Winning Doesn't Always Equal Success," filmed December 6, 2019, at TEDWomen2019, Palm Springs, CA, TED video, 15:40, https://www.ted.com/talks/valorie_kondos_field_why_winning_doesn_t_always_equal_success.

17 Field, "Why Winning."

18 Field, "Why Winning."

19 Field, "Why Winning."

20 Field, "Why Winning."

21 This concept was influenced by Marcus Buckingham, "What Great Managers Do," Harvard Business Review, March 2005, https:// hbr.org/2005/03/what-great-managers-do.

22 Anita Williams Woolley et al., "Evidence for a Collective Intelligence Factor in the Performance of Human Groups," Science 330, no. 6004 (2010): 688, https://doi.org/10.1126/science.1193147.

23 Anita Woolley and Thomas W. Malone, "Defend Your Research: What Makes a Team Smarter? More Women," Harvard Business Review, June 2011, https://hbr.org/2011/06/ defend-your-research-what-makes-a-team-smarter-more-women.

24 Peter Gwynne, "Group Intelligence, Teamwork, and Productivity," Research-Technology Management 55, no. 2 (March–April 2012): 7–8, https://www.jstor.org/stable/26586213.

25 Kevin M. Kniffin et al., "Eating Together at the Firehouse: How Workplace Commensality Relates to the Performance of Firefighters," Human Performance 28, no. 4 (2015): 281–306, https://doi.org/10.1080/08959285.2015.1021049.

26 Tami Kim et al., "Work Group Rituals Enhance the Meaning of Work," Organizational Behavior and Human Decision Processes 165 (July 2021): 197–212, https://doi.org/10.1016/j. obhdp.2021.05.005.

27 Julia Rozovsky, "How Google Thinks about Team Effectiveness," re:Work (blog), October 25, 2017, https://rework.withgoogle.com/blog/how-google-thinks-team-effectiveness/.

28 Rozovsky, "How Google Thinks."

29 Amy C. Edmondson, The Fearless Organization: Creating Psychological Safety in the Workplace for Learning, Innovation, and Growth (Hoboken, NJ: John Wiley & Sons, 2019), xvii.

30 Amy Edmondson, "Psychological Safety and Learning Behavior in Work Teams," Administrative Science Quarterly 44, no. 2 (1999): 354, https://doi.org/10.2307/2666999.

31 Nigel Bassett-Jones, "The Paradox of Diversity Management, Creativity and Innovation," Creativity and Innovation Management 14, no. 2 (June 2005): 169–175, https://doi.org/10.1111/j.1467-8691.00337.x.

32 Henrik Bresman and Amy C. Edmondson, "Research: To Excel, Diverse Teams Need Psychological Safety," Harvard Business Review, March 17, 2022, https://hbr.org/2022/03/research-to-excel-diverse-teams-need-psychological-safety.

33 Bryce G. Hoffman, American Icon: Alan Mulally and the Fight to Save Ford Motor Company (New York: Crown Business, 2012), 92; Alan Mulally, discussion with the author, July 2022.

34 If you want to learn more about Mulally's "Working Together" leadership and management system, check out the following resources: Frances Hesselbein, Marshall Goldsmith, and Sarah McArthur, eds., Work Is Love Made Visible: A Collection of Essays about the Power of Finding Your Purpose from the World's Greatest Thought Leaders (Hoboken, NJ: John Wiley & Sons, 2019); Alan Mulally and Sarah McArthur, "A Conversation with Alan Mulally about His 'Working Together' Strategic, Operational, and Stakeholder-Centered Management System," Leader to Leader 2022, no. 104 (Spring 2022): 7–14, https://doi.org/10.1002/ltl.20628.

35 Alan Mulally, discussion with the author, July 2022.

36 Tasha Eurich, Insight: Why We're Not as Self-Aware as We Think, and How Seeing Ourselves Clearly Helps Us Succeed at Work and in Life (New York: Crown Business, 2017), 219.

37 Alan Mulally, discussion with the author, July 2022.

38 John Schaubroeck, Simon S. K. Lam, and Ann Chunyan Peng, "Cognition-Based and Affect-Based Trust as Mediators of Leader Behavior Influences on Team Performance," Journal of Applied Psychology 96, no. 4 (2011): 863–871, https://doi.org/10.1037/a0022625.

39 Paul J. Zak, "The Neuroscience of Trust," Harvard Business Review, January–February 2017, https://hbr.org/2017/01/the-neuroscience-of-trust.

40 Paul J. Zak, The Trust Factor: The Science of Creating High-Performance Companies (New York: AMACOM, 2017), 16–17.

41 Paul J. Zak, Robert Kurzban, and William T. Matzner, "Oxytocin Is Associated with Human Trustworthiness," Hormones and Behavior 48, no. 5 (December 2005): 522-527, https://doi.org/10.1016/j.yhbeh.2005.07.009.

42 Christine Porath, Mastering Community: The Surprising Ways Coming Together Moves Us from Surviving to Thriving (New York: Balance, 2022), 18.

43 Jonathan D. Rockoff, "Celebrating Failure in a Tough Drug Industry," Wall Street Journal, March 3, 2017, https://www.wsj.com/articles/celebrating-failure-in-a-tough-drug-industry-1488568710.

44 Tami Kim et al., "Work Group Rituals Enhance the Meaning of Work," Organizational Behavior and Human Decision Processes 165 (July 2021): 197–212, https://doi.org/10.1016/j.obhdp.2021.05.005.

45 Giada Di Stefano et al., "Learning by Thinking: How Reflection Improves Performance" (HBS Working Paper Number 14-093, Harvard Business School, Boston, MA, March 2014), http://dx.doi.org/10.2139/ssrn.2414478.

46 Joann S. Lublin, "Power Couple: Meet the Sister CEOs," Wall Street Journal, last modified July 18, 2011, https://www.wsj.com/articles/SB10001424052702304223804576446173524792108.

47 Boris Groysberg, Sarah L. Abbott, and Robin Abrahams, "Maggie Wilderotter: The Evolution of an Executive," Harvard Business School, Jun 30, 2017, https://hbsp.harvard.edu/product/417091-PDF-ENG.

48 Ben Horowitz, What You Do Is Who You Are: How to Create Your Business Culture (New York: HarperBusiness, 2019), 168.

49 Horowitz, What You Do, 170.

50 The Cambridge Cyber Summit, "Maggie Wilderotter," CNBC, last modified October 14, 2017, https://www.cnbc.com/2016/09/08/maggie-wilderotter.html.

51 Christine Porath, "Half of Employees Don't Feel Respected by Their Bosses," Harvard Business Review, November 19, 2014, https://hbr.org/2014/06/the-power-of-meeting-your-employees-needs.

52 Christine Porath, Mastering Community: The Surprising Ways Coming Together Moves Us from Surviving to Thriving (New York: Balance, 2022), 58.

53 Porath, "Half of Employees."

54 Christine Porath, Mastering Civility: A Manifesto for the Workplace (New York: Grand Central Publishing, 2016), 17.

55 Arieh Riskin et al., "The Impact of Rudeness on Medical Team Performance: A Randomized Trial," Pediatrics 136, no. 3 (September 2015): 487–495, https://doi.org/10.1542/peds.2015-1385.

56 Sue Shellenbarger, "Tuning In: Improving Your Listening Skills," Wall Street Journal, last modified July 22, 2014, https://online.wsj.com/articles/tuning-in-how-to-listen-better-1406070727.

57 Jane M. Von Bergen, "Helping Accountants Find a Higher Purpose at KPMG," Philadelphia Inquirer, June 13, 2017, https://www.inquirer.com/philly/business/labor_and_unions/finding-higher-purpose-work-kpmg-pfau-retention-turnover-engagement-accounting-healthcare-20170613.html.

58 Ewin Hannan, "How KPMG Gave 6000 Employees a Higher Purpose," Australian Financial Review, May 6, 2016, https://www.afr.com/work-and-careers/management/how-kpmg-gave-6000-employees-a-higher-purpose-20160427-gog7sa.

59 Bruce N. Pfau, "How an Accounting Firm Convinced Its Employees They Could Change the World," Harvard Business Review, October 6, 2015, https://hbr.org/2015/10/how-an-accounting-firm-convinced-its-employees-they-could-change-the-world.

60 Adam M. Grant, "The Significance of Task Significance: Job Performance Effects, Relational Mechanisms, and Boundary

Conditions," Journal of Applied Psychology 93, no. 1 (2008): 108–124, https://doi.org/10.1037/0021-9010.93.1.108.

61 Shawn Achor, Big Potential: How Transforming the Pursuit of Success Raises Our Achievement, Happiness, and Well-Being (New York: Currency, 2018), 87–91.

62 Robert Pearl, "What Health Systems, Hospitals, and Physicians Need to Know about Implementing Electronic Health Records," Harvard Business Review, June 15, 2017, https://hbr.org/2017/06/what-health-systems-hospitals-and-physicians-need-to-know-about-implementing-electronic-health-records.

63 Dan Ariely, Emir Kamenica, and Dražen Prelec, "Man's Search for Meaning: The Case of Legos," Journal of Economic Behavior & Organization 67, no. 3–4 (September 2008): 671–677, https://doi.org/10.1016/j.jebo.2008.01.004.

64 Adam M. Grant and Francesca Gino, "A Little Thanks Goes a Long Way: Explaining Why Gratitude Expressions Motivate Prosocial Behavior," Journal of Personality and Social Psychology 98, no. 6 (2010): 946–955, https://doi.org/10.1037/a0017935.

65 Adam M. Grant, "The Significance of Task Significance: Job Performance Effects, Relational Mechanisms, and Boundary Conditions," Journal of Applied Psychology 93, no. 1 (2008): 108–124, https://doi.org/10.1037/0021-9010.93.1.108.

66 All quotes and details from this story are from the following: Jesse Cole, discussion with the author, June 2022.

67 Arne Gast et al., "Purpose: Shifting from Why to How," McKinsey Quarterly, April 22, 2020, https://www.mckinsey.com/business-functions/people-and-organizational-performance/our-insights/purpose-shifting-from-why-to-how.

68 Christopher Michaelson, Douglas A. Lepisto, and Michael G. Pratt, "Why Corporate Purpose Statements Often Miss Their Mark," strategy+business, August 17, 2020, https://www.strategy-business.com/article/Why-corporate-purpose-statements-often-miss-their-mark.

69 Jim Harter, "U.S. Employee Engagement Slump Continues," Gallup, April 25, 2022, https://www.gallup.com/workplace/391922/employee-engagement-slump-continues.aspx.

70 Adam Grant, "How Customers Can Rally Your Troops," Harvard Business Review, June 2011, https://hbr.org/2011/06/how-customers-can-rally-your-troops.

71 Adam M. Grant and David A. Hofmann, "It's Not All about Me: Motivating Hand Hygiene among Health Care Professionals by Focusing on Patients," Psychological Science 22, no. 12 (2011): 1494–1499, https://doi.org/10.1177/0956797611419172.

72 Claudine Gartenberg, Andrea Prat, and George Serafeim, "Corporate Purpose and Financial Performance," Organization Science 30, no. 1 (January–February 2019): 1–18, https://doi.org/10.1287/orsc.2018.1230.

73 William C. Taylor, Simply Brilliant: How Great Organizations Do Ordinary Things in Extraordinary Ways (New York: Portfolio/Penguin, 2016), 106.

74 Adrian Gostick and Chester Elton, Leading with Gratitude: Eight Leadership Practices for Extraordinary Business Results (New York: Harper Business, 2020), 115.

75 Garry Ridge, Tribe Culture: How It Shaped WD-40 Company (Dublin, OH: Telemachus Press, 2020), 49.

76 Gostick and Elton, Leading with Gratitude, 104.

77 Garry Ridge, discussion with the author, July 2022.

78 Garry Ridge, discussion with the author, July 2022.

ABOUT THE AUTHOR

DR. DAVID BURKUS IS ONE OF THE WORLD'S LEADING business thinkers. His forward-thinking ideas and best-selling books are helping leaders and teams do their best work ever.

He is the bestselling author of four books about business and leadership. His books have won multiple awards and been translated into dozens of languages. His insights on leadership and teamwork have been featured in the *Wall Street Journal*, *Harvard Business Review*, *USAToday*, *Fast Company*, the *Financial Times*, *Bloomberg BusinessWeek*, CNN, the BBC, NPR, and CBS This Morning. Since 2017, Burkus has been ranked multiple times as one of the world's top business thought leaders. As a sought-after international speaker, his TED Talk has been viewed over two million times. He has worked with leaders from organizations across all industries, including PepsiCo, Fidelity, Clorox, Adobe, and even NASA.

A former business school professor, Burkus holds a master's degree in organizational psychology from the University of Oklahoma, and a doctorate in strategic leadership from Regent University.

He lives outside of Tulsa with his wife and their two boys.

Printed in the USA
CPSIA information can be obtained
at www.ICGtesting.com
LVHW100323150823
755208LV00004B/410